T0091932

REEDS MARINE ENGINEERING AND TECHNOLOGY

SHIP CONSTRUCTION
FOR MARINE ENGINEERS

REEDS MARINE ENGINEERING AND TECHNOLOGY SERIES

5

REEDS MARINE ENGINEERING AND TECHNOLOGY

SHIP CONSTRUCTION
FOR MARINE ENGINEERS

7TH EDITION

Revised by Paul A Russell

E A Stokoe

REEDS

LONDON • OXFORD • NEW YORK • NEW DELHI • SYDNEY

REEDS
Bloomsbury Publishing Plc
50 Bedford Square, London, WC1B 3DP, UK
29 Earlsfort Terrace, Dublin 2, Ireland

BLOOMSBURY, REEDS and the Reeds logo are trademarks
of Bloomsbury Publishing Plc

First edition published by Thomas Reed Publications 1964
Second edition 1968
Third edition 1973
Fourth edition 1979
Fifth edition 1979
Sixth edition published by Adlard Coles Nautical 2016
Reissued 2019
This edition published 2022

Copyright © Paul Anthony Russell and Adlard Coles Nautical 2016, 2022

Paul Anthony Russell has asserted his right under the Copyright, Designs and
Patents Act, 1988, to be identified as Author of this work.

All rights reserved. No part of this publication may be reproduced or
transmitted in any form or by any means, electronic or mechanical, including
photocopying, recording, or any information storage or retrieval system,
without prior permission in writing from the publishers.

No responsibility for loss caused to any individual or organization
acting on or refraining from action as a result of the material in
this publication can be accepted by Bloomsbury or the author.

A catalogue record for this book is available from the British Library.

Library of Congress Cataloguing-in-Publication data has been applied for.

ISBN: PB: 978-1-4729-8920-8
ePDF: 978-1-4729-8921-5
ePub: 978-1-4729-8919-2

2 4 6 8 10 9 7 5 3 1

Typeset in Myriad Pro 10/14 by Newgen Knowledge Works (P) Ltd., Chennai, India
Printed and bound in Great Britain by CPI Group (UK) Ltd, Croydon CR0 4YY

To find out more about our authors and books visit
www.bloomsbury.com and sign up for our newsletters.

CONTENTS

PREFACE

The update of this publication comes at such an exciting time for International Shipping. Material Science is marching on at a great rate of knots, and the regulators are continually looking at ways to ensure that ships are built to a specification that is safer than before. This volume is designed to provide knowledge primarily about the construction of ships engaged in international voyages. The content is intended to cover the requirements of the various Flag Administration qualifications for marine engineers studying for the International Maritime Organization's (IMO) Standards of Training, Certification & Watchkeeping (STCW) for seafarers at the Certificates of Competency as Chief Engineering Officer, Second Engineering Officer and Engineering Officer of the Watch (EOOW) levels.

The publication complements *Reeds Vol 4*, *Vol 8* and *Vol 12* of this series. It will also be found useful by those studying for STCW Chief Mate and Master's Examinations as well as students undertaking maritime degrees and Superyacht Qualifications.

The book will also cover the subject to the level required by people starting out or working in the operation, repair and surveying of ships. It is intended to give an indication about the typical methods of construction, and it is suggested that engineers studying at sea should compare the arrangements shown in the book with those on the ship wherever possible. In this way students will relate the descriptions in the book to the actual structure of ships in preparation for the Flag Administration's Examinations. The typical examination questions are intended as a revision of the whole work.

In addition, the classification society's 'rules' for the construction of ships are very complex and comprehensive and also available to inspect. This publication is designed to give an overview of the subject enabling the student to better comprehend the detail of the classification society's rules.

GENERAL

INTRODUCTION

International Shipping started with the needs of manufacturers to transport raw materials from suppliers and finished goods to customers. As business grew so did the distances that needed to be covered.

Traditional seafaring nations built up merchant fleets that would carry goods to and from any place that was required. No longer bound by national boundaries, ship owners would seek work for their ships on a global basis. Therefore, ship owners not only traded their vessels to and from the country of registration but they would also carry cargo to and from any other country where business was to be found. The rules for the construction and use of these vessels were set by the nation where the vessel was registered.

These nations built up considerable expertise relating to the safety of the vessels falling within their administration. Their rules covered the construction and use of the vessel as well as the competence of staff. However, the shipping industry is a global business and during the 1970s owners found that other nations were becoming interested in registering and setting the rules for operating ships on international voyages.

As the industry looked for different nations to provide the registration of their ships there was a growing need to set standards that all flag administrations could uphold. The organisation with the necessary infrastructure for completing this task was the United Nations (UN). The International Maritime Organization (IMO) is the UN's specialist agency concentrating on maritime matters.

The IMO secretariat provides the structure for the contracting member nations to record their businesses, ideas and decisions. It has its headquarters in London (UK) and has the remit for the development of all the rules and regulations that govern International Shipping. These rules cover the technical development of ships, design and specification of equipment used on board, fire protection, safety of navigation, radio communication, search and rescue, training and certification of watchkeepers, carriage of cargoes, port and flag state responsibilities and international security relating to the maritime industry. More about the rules, regulations and standards associated with ships can be found in Chapter 10.

It is the member states of the IMO that complete the decision making process within the Agency; however they also have non-government organisations (NGOs) that provide the technical support and guidance when necessary.

The technical guidance relating to the strength, safe operation and correct and efficient design of a ship and its associated systems comes from naval architects, marine engineers and navigation experts.

These specialists work and operate through organisations such as classification societies, the professional institutes and other organisations representing different sectors of the industry such as the 'oil' sector or 'ship masters'.

Over the past few decades there have been significant improvements in ship design and construction. Much more is known about material science, including the properties and performance characteristics of different metals, alloys and composites. Quality assurance runs through the whole process. If the manufacturing process of the raw materials is correct and audited and the construction is closely monitored, then the finished vessel should perform to a known design specification.

This used to be the case in the past of course; however scantlings had to be larger and construction more complex due to the uncertainties in the performance of detailed parts and lack of a complete understanding about how they interacted with other parts of the ship. Using modern computerised design techniques, the stresses in individual components can be calculated in a way that was not possible just a few decades ago. Classification societies and other design companies employ 'Naval Architects' who have significant experience in this area and are very important for providing the industry with the constructional details necessary for building ships. More can be studied about the work of these organisations in Chapter 10.

Ship design starts with the need to transport either raw materials or manufactured goods from one part of the world to another via the water route. It follows on to also capture the need to transport people and vehicles from place to place. If the vessel is to spend its working life in sheltered inshore waters or rivers, it might be built to a different set of construction rules than a vessel that is built to sail the world's oceans.

Ships that are expected to travel to areas where ice is present will need to be constructed to an ice class standard. The strategic importance of the Arctic is growing all the time as the world's temperatures seem to be rising and the need for the construction of ships, with an ice class rating, is growing.

Owners are looking to take advantage of sailing in the polar regions of the world, and for this reason the IMO has developed constructional and operational guidelines for such vessels. There are seven polar classes for which the International Association of Classification Societies (IACS) has developed constructions rules and the IMO has included additional requirements for vessels operating to the 'International Code for Ships Operating in Polar Waters' (the Polar Code).

The ship's owner's will also need to specify further criteria for their ship, which will be determined by their business plan. This criteria will include any width, draught and height restrictions for the ports or routes where the vessel will be expected to travel. Types and volumes of cargo to be carried and cargo handling equipment will also need to be considered, as well as the speed required and length of service. The cost of the vessel will be very important, and this will determine the standard of equipment to be fitted.

Having said that it is also fair to say that most new tonnage is being built to high standards and some of the equipment is at the cutting edge of technology. Marine propulsion motors, for example, cannot be built more powerful as material science is the limiting factor.

The maritime industry is focused on improving their environmental footprint. This brief reaches out to every part of the process for the design and construction of ships. Saving fuel is a step in the right direction and saves the owners money. New ideas in design will contribute to this characteristic and some of these changes will be due to the increased constructional ability.

Keeping the ship's staff and passengers safe is always a top priority for everyone including designers, construction craftsmen and operational staff. All ships will be built to a set of rules, determined by the classification society responsible for overseeing the construction of the vessel.

Class rules will be developed from experience and from the requirements of the regulators. For example, the IMO require that all ships are built to the Safety of Life at Sea (SOLAS) requirements. However, passenger ships must also be designed to the Safe Return to Port (SRtP) requirements.

A further recent development has been the view of calculating damage stability. Again the development in computer technology means that the designers can better calculate the survivability of a vessel that has been damaged. This will allow the internal compartments to be designed in such a way as to give the vessel the best chance of survival following any damage to the outside hull.

The basic ship types and associated terms and nomenclature appear in Chapter 1 and further design selection criteria can be found in Chapter 12.

1

SHIP TYPES AND TERMS

The design and construction of merchant ships are driven initially by commercial considerations as well as by the requirement to build safe machinery. Therefore, ships vary considerably in size, type, layout and function. In broad categories, modern tonnage can be divided into passenger ships, cargo ships and specialised vessels built for specific types of work. This book deals with the general construction details of these different ship categories.

Cargo ships may be further subdivided into those designed to carry liquid cargoes, those designed to carry non-liquid cargoes and those intended to carry specialised cargoes such as gas and chemical carriers.

The larger cargo ships (over 15 000 gt) are designed to carry as much cargo, in one voyage, as possible. They usually sail between significantly sized 'hub' ports that have been designed to accommodate their size.

As ships are now so big, care must be taken when considering the passage that they will undertake as part of their required service. For example, a 250 000 gt fully laden supertanker will, most likely, not be able to sail through the Suez Canal, and will instead be forced to sail around the tip of Africa when on passage from the Middle East to Europe.

However, if an owner wishes to build a vessel that is just big enough to transit the Suez Canal, then that vessel will attract the term 'Suez Max'. A similar situation exists with the Panama Canal. The largest vessel that is able to transit this canal will be called a 'Panamax' sized vessel.

This procedure of labelling ships has gained momentum, and students may, for example, hear terms such as 'Dunkerque Max' being used to describe the size of a ship.

Passenger Ships

During the first half of the twentieth century the *passenger ship* was the only way to transport large numbers of people about the world reliably. However, the development of air transport took over this role, and the *passenger ship* market changed almost overnight. The vast majority of ships in this sector are now cruise ships and ferries. Most are of a conventional design, but the *fast ferry* is very popular for short service routes.

Ferries operating in a limited area or on a short regular route are increasingly being designed using liquefied natural gas (LNG) or 'hybrid' technology as the energy source for vessel propulsion.

Universal agreement has the *passenger ship* defined as 'a vessel that is designed to carry more than 12 passengers on an international voyage'. These vessels must comply with the relevant International Maritime Organization (IMO) regulations included in the Safety Of Life At Sea (SOLAS) and the Load Line Conventions for passenger ships. Cargo ships can still carry up to 12 passengers without being reclassed as a passenger ship.

The only departure from this requirement is in the superyacht sector, where the UK's Maritime Coastguard Agency (MCA) has developed the 13–36 Passenger Yacht Code. This code is for use by the Red Ensign Group to register large passenger yachts that carry up to 36 passengers. These vessels find it very difficult to comply with the full requirements of the IMO regulations for passenger ships, which led the UK administration to develop the regulations for this sector.

Passenger ships, in general, range from small river ferries to large ocean-going vessels. Cruise ships can now carry over 8000 passengers and are designed to provide maximum comfort for all guests on board. These ships include in their services large dining rooms, luxury restaurants, theatres, cinemas, swimming pools with water slides, gymnasia, open deck spaces and shops. They usually cater for guests with a range of purchasing power and are being designed increasingly to provide the majority of rooms with a balcony and a 'sea view'.

Ferries are also being designed after listening to customers' needs. For example, ferries carrying large numbers of trucks might provide a specific place where the drivers can rest and have a meal in a restaurant that has been designed with them in mind. Roll-on/roll-off (Ro-Ro) ferries are now 'double decked' and arranged so that both decks can be loaded at the same time. This is important as the time in port for any vessel is non-revenue-earning and therefore needs to be reduced to a minimum.

Small 'passenger vessels' are now on the increase, some supplying the ever-growing needs of the 'offshore' renewable energy sector. Passenger vessels restricted to 'inland waterways' are starting to use 'hybrid' technology to meet their propulsion needs. There is an advantage for passenger comfort if electric engines can be used. However, there will have to be a significant development in battery technology, and/or super capacitors, before the diesel engine can be removed from the system completely from all but the smallest of vessels.

Invariably new standards are applied to passenger ships slightly ahead of cargo ships. It is also usually the case that the rules are more stringent for passenger ships, and during the 1990s there was increasing concern about the safety of very large passenger ships and new regulations appeared focusing on their construction and on the training of the crew for such vessels.

Minimum standards for crew accommodation are now required under the International Labour Organization's (ILO) Maritime Labour Convention 2006 (MLC 2006).

A sign of the times is that most people look to manage risk as they go about their daily work. The IMO is no different and during the 1990s and into the early twentieth century they worked on developing a risk-based approach to the operation and construction of ships.

SOLAS Chapter II-1. This section of SOLAS relates to the structure, subdivision and stability, machinery and electrical installations for passenger and cargo ships. This part of SOLAS is continually being updated in light of experience and to improve the survivability of ships in the event of damage.

Significant updates came in the 2009 version, with further updates introduced for passenger ships built after 2020. These all relate to the method of calculating the stability of a ship and constructing the internal subdivisions to give it the best chance of survival in the event of damage to the hull.

Also, in 2006 came the plans to improve the safety of passenger ships based around the Safe Return to Port (SRtP) concept for these ships. SRtP centres around the notion that 'your ship is your best lifeboat', and therefore if the ship can be constructed with maximum 'survivability', then there will be less of a need to resort to the much smaller and more vulnerable 'lifeboats' in the event of an accident.

The IMO amended regulations to improve the safety of passenger ships by placing more emphasis on the prevention of casualties and improving the survivability of the ship in the event of an incident, and thus allowing the passengers to stay safely on board while the vessel returns to the nearest safe haven.

A further development for the very large passenger ship is the 'diesel electric' (power station) concept. This is where the main engines are large diesel alternators producing high voltage electricity that is used to either power the ship or run large electrical loads servicing the passengers, such as the air conditioning compressors, ventilation fans or galley and laundry equipment.

Two marine manufacturers are now making large electric motors that are arranged in their own containers (pods) and hang below the hull of the ship. See page 136, Figure 8.15 for more detail about their arrangement.

Cruise ships that can carry more than 8000 people are now being built, and in the event of a major disaster, those people may need evacuating from the vessel. SOLAS sets out the rules for calculating the minimum number and maximum capacity for the lifeboat requirements for each vessel (see Figure 1.1).

SOLAS regulations set out the general requirements for lifeboats and state that no lifeboat shall be approved with a passenger capacity of more than 150 persons. However, on a vessel carrying more than 8000 people, larger lifeboats have been developed that require special permission from the flag state to enable them to be used as they are outside of the current regulations.

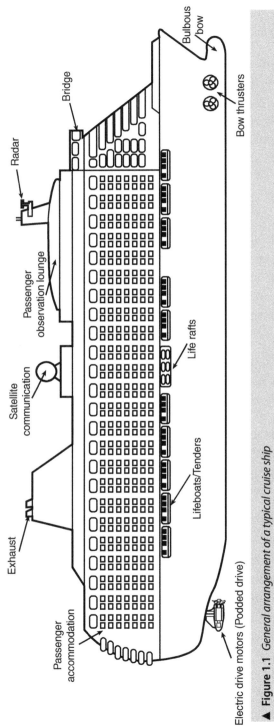

▲ **Figure 1.1** *General arrangement of a typical cruise ship*

Container Ships (the Modern Cargo Liner Ships)

Ships travelling between allocated ports and having specific departure and arrival dates/times are considered to be running on a *liner* trade. Both passenger ships and cargo ships can be 'liner ships', however the role of carrying and delivering goods and services has now moved mostly to the domain of the cargo ship with the container vessel being the modern version.

Cargo liners are vessels designed to carry a variety of cargoes between specific ports and, as stated, the modern configuration of this type of vessel is the *container ship* and most of the non-liquid cargoes and some liquid cargoes are now carried around the world by this type of ship.

The standardisation of cargo carrying 'units' (containers) has transformed the efficiency of moving cargo from place to place. Anything that will fit into a standard (8 ft x 8 ft 6 in x 20 ft or 40 ft) sized container can be moved via an intermodal transportation system in which the container ship plays a central role in the waterborne part of that system.

The reusable containers can be insulated and refrigerated and are capable of carrying perishable cargoes such as meat, fruit and fish. They are mostly standard steel boxes with doors at one end and are used to transport goods that are packed in boxes, drums, bags and cases or stacked on pallets. More details about containers can be found in Chapter 9 of this book.

Figure 1.2 shows the layout of a modern *container ship*, the size of which can be measured by the number of containers it can carry. The sizing is in the form of twenty-foot equivalent (TEU) units. This means that a medium sized container ship could be described as having a size of 9000 TEU or as having the ability to carry 9000 twenty-foot containers. The actual containers carried could be a mixture of twenty foot and forty foot.

The global system is arranged so that there are large 'deep water' ports, known as 'hub ports', that are distributed about the world. Large container ships up to 18 000 TEU travel between these ports. Smaller vessels known as 'feeder' vessels move the cargo from the large ports to smaller ports, that are close to the hub port.

As with many ships, at the extreme forward end is a tank known as the *fore peak* which may be used to carry water ballast or fresh water. Above this tank is an area called the *chain locker* and also a *storage space*. At the after end is a tank known as the *after peak* which generally encloses the stern tube in a watertight compartment. At the bottom of the vessel and between the two peak bulkheads is a continuous tank top forming a

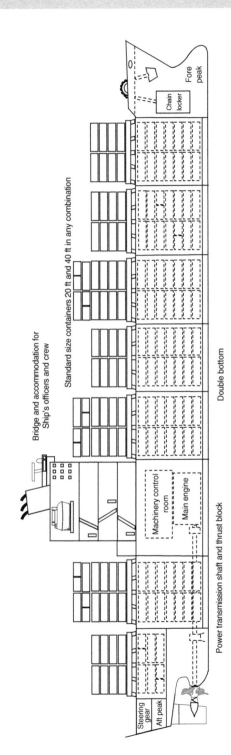

▲ **Figure 1.2** *Container ship*

double bottom space which is further subdivided into smaller tanks suitable for carrying oil fuel, fresh water and water ballast.

The *machinery space* consists of heavy equipment and as such will place considerable 'local' stresses on the structure of the ship. If placed at the aft end, as with Figure 1.3, the machinery will exert a maximum bending moment on the ship's hull. If placed in a more central position, as shown in Figure 1.2, then the 'light ship' bending moment and hull stress will be reduced. The latest diesel electric propulsion systems enable the designer to place the main diesels in the best possible place for the benefit of the hull thereby maximising the cargo carrying capacity.

The reason for this is that the positions of the engines are not determined by the propeller shafting and gearboxes. The connection to the inboard electric motor or podded drive will be via electric cables and not by mechanical equipment, such as power transmission shafts and gearboxes.

Currently, in 2022, most ships use some form of oil as their major energy source. The general term for this fuel is *bunkers*, which is a legacy from the early days of shipping when coal was the main source of fuel and the coal was loaded into *coal bunkers*. The *oil fuel bunkers* are taken on board when required and loaded into designated tanks called *bunker tanks*.

Due to the possibility of the presence of water, the fuel is transferred into *settling tanks* where any water in the mix will settle to the bottom of the tank. The water can then be drained off prior to the fuel being further treated on board with filters and centrifugal purifiers before being used in the engines. In order to help the stability of a ship, it is generally a good idea to have the bunker tanks as low in the ship as practical, and therefore the tanks used as bunker tanks are usually the double bottom tanks. (See *Reeds Vol 8* for more information about fuel tanks, storage and treatment equipment.)

It is also advantageous to have the bunkers arranged as close to the machinery space as possible. Fuel oil storage has become a significant issue in the design of ships. This is due to the requirement to have several different types or grades of fuel stored separately from each other.

In the past this has meant that the bunker tanks could be very close to or in contact with the outer hull of the ship. This ensured that any breach of the hull with respect to the tank would release oil into the sea. The MARPOL (International Convention for the Prevention of Pollution for Ships) regulation developed in 2006 and brought into force during 2010 required new builds to have 'protected' fuel oil storage tanks when an individual tank holds more than 600 m^3 of fuel.

When ships have the engine room sighted around the centre part of the vessel, there is a long path for the propeller shaft to take before it exits the hull via the stern tube bearing.

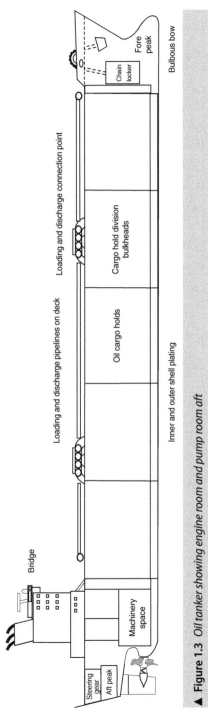

▲ **Figure 1.3** *Oil tanker showing engine room and pump room aft*

Between the aft engine room bulkhead and the after peak bulkhead is a watertight *shaft tunnel* enclosing the shaft and allowing access to the intermediate shaft and bearings directly from the engine room. An exit in the form of a vertical trunk is arranged at the after end of the tunnel in case of emergency. In a twin screw ship it may be necessary to construct two such tunnels, although they may be joined together at the fore and aft ends.

The arrangement of the machinery space on many modern ships is further aft and therefore the propeller shafting and intermediate bearings are sighted in the machinery space. Invariably a walkway and guard rail are placed close to the rotating shaft so that the watch keeping engineer can safely inspect the drive train for correct operation at any time.

The MSC, at its 87th session in May 2010, adopted a new SOLAS regulation II-1/3–10 on goal based ship construction standards for bulk carriers and oil tankers (resolution MSC.290(87)).

These rules require that vessels have a design life of at least 25 years and that all the structural and manufacturing rules are consistent with this requirement. The class rules should consider at least:

- extreme loads
- design loads
- fatigue
- corrosion.

General Cargo Ships

General cargo ships used to be known as 'tramp' ships, as well as 'liner' ships. The liner trade was the description given to ships that were on a regular trading route. The ships would call at the same ports on a preset time scale. This trade pattern has now been taken over by container ships.

The ships that were used to carry any specific type of cargo and travel anywhere in the world on an irregular route were known to be trading as tramp ships. They were often hired out on a spot or time charter to carry bulk cargo or general cargo, and were usually slower and smaller than the container 'liner' vessels. The modern general cargo vessels may still carry some or all of their cargo in containers. The general arrangement is shown in Figure 1.4.

To assist the safe stowage of cargo the cargo space is divided into lower *holds* and compartments between the decks, or *'tween* decks. Many ships have three decks,

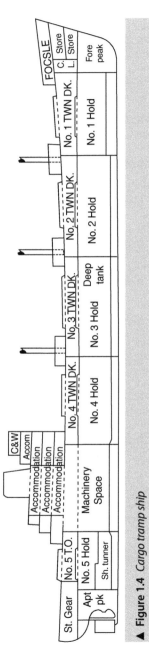

▲ **Figure 1.4** *Cargo tramp ship*

thus forming upper and lower 'tween decks. This system allows different cargoes to be carried in different compartments and reduces the possibility of the cargo getting crushed. Access to the cargo compartments is provided by means of large hatchways which are now all closed by hydraulically operated steel hatch covers (see Chapter 11).

Some of these ships will be refrigerated, giving it the ability to carry fruit and other 'perishable goods' such as fish and sometimes meat.

The insulation used in the cargo holds of reefer ships is explained in Chapter 12. However, the general arrangement of the reefer ship is very similar to the example of the general cargo vessel shown in Figure 1.4.

The difference would be that the holds and hatch covers will be insulated and additional refrigeration equipment will be required to reduce the temperature in the hold and to keep the temperature low for the duration of the voyage.

For example, a hold full of bananas will need to be loaded as quickly as possible, and then as soon as the hatch covers are closed, the hold temperature will need to be reduced to 4°C +/− 0.1°C within 48hrs of loading the cargo.

The cargo of bananas is carried at this temperature for the duration of the voyage. To achieve this the machinery must be working to the peak of its ability. Therefore, for this to happen, the ship's staff must ensure that all the equipment's maintenance is up to date. The opportunity to carry out this work would be during the ballast passage.

Suitable cargo handling equipment is provided in the form of hydraulic or electrically powered cranes. Heavy lift equipment may be fitted and is usually situated next to one or more hatches. A *forecastle* is fitted to reduce the amount of water shipped forward and to provide adequate working space for handling ropes and cables.

The *forecastle* also acts very effectively to protect the forward hatches from heavy weather damage. Hatch covers are a prime area to guard against a breach of watertight integrity. The hatch covers are well designed, but the *forecastle* provides a very good first line of defence and deflects the full force away from the hatches. See Chapter 5 for more details on the arrangement of hatch covers.

However, the work of these vessels is now being taken over by bulk carriers and smaller container vessels. Figure 1.4 shows the layout of a typical cargo tramp ship. The significant advantage of these vessels is 'flexibility' so a large variety of cargo can be carried. The space immediately forward of the machinery space may be subdivided into lower 'tween decks and *hold/deep tank*, thus improving the ability to even out the stress on the hull and/or give different options for carrying the different cargo and/or liquids such as fuel or water or dangerous goods. To contribute to the flexibility of these vessels they are fitted with cargo handling equipment, such as cranes, which replace the older cargo handling equipment called derricks.

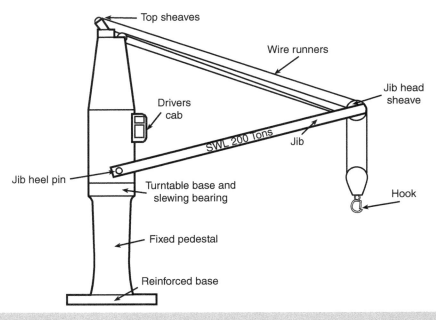

▲ **Figure 1.5** *Cargo crane*

Roll-On/Roll-Off Vessels

These vessels are designed with flat decks and have moveable watertight divisions to enable vehicles and tractor-trailer units to be driven into and off the vessel under their own power. Having such a large continuous deck means that any appreciable accumulation of water will have a magnified adverse effect on the vessel's stability. This is due to the 'free surface effect' inherent in such a body of water. (See *Reeds Vol 4*, Chapter 5). The Ro-Ro vessel is particularly susceptible to this feature, and all the staff should be well aware of the dangers of the ship becoming unstable if it is not operated correctly.

A ramp is fitted at one or at both ends of the ship allowing direct access for cars, trucks and buses which remain on board in their laden state. These ramps lead to large outer doors which have in the past been a source of leakage due to damage and/or incorrect operation. Any ingress of water will come straight onto the Ro-Ro decks leading to the possible problem with free surface effect mentioned earlier.

Containers may be loaded 2 or 3 high by means of forklift trucks. Lifts and inter-deck ramps are used to transfer vehicles between decks. Most modern vessels have stern ramps that are angled to allow vehicles to be loaded from a straight quay (Figure 1.6). This would circumvent the need for a special 'link span' arrangement or blind end berth.

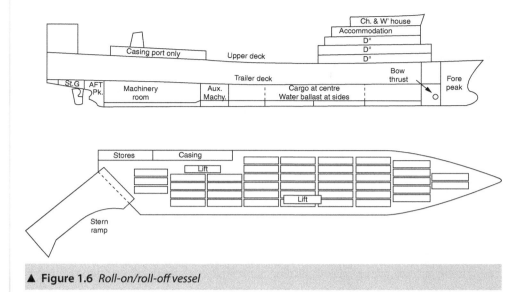

▲ **Figure 1.6** *Roll-on/roll-off vessel*

Some specialist Ro-Ro vessels have very large car carrying capacity and are used to move considerable numbers of new cars around the world. Loading these weights (containers) will quickly alter the trim of the vessel and move it away from the vertical position. This will have the effect of not being able to drop subsequent containers vertically into position. Therefore, there must be a means of keeping the ship upright. This is achieved by pumping water from one side of the vessel to the other. The movement of the water weight counterbalances the added weight of the containers (or cars) and is the method used to achieve the requirement of keeping the vessel vertical. The name of the system is called the 'heeling arrangements' and more details can be seen on page 199.

Where the vessel has a combined container/Ro-Ro capacity the term *Lo-Lo* is sometimes used. This refers to the lift-on lift-off feature of the containerised cargo.

Accommodation for crew and passengers includes restaurants and fast food outlets as well as specific areas for long distance lorry drivers where they can complete 'official' rest periods that satisfies the 'drivers' hours' regulations. Again these vessels tend to work as liner vessels and ferries. The ports of Dover and Calais, either side of the English Channel, are very good examples of ports that are highly specialised in handling the Ro-Ro ferry operation.

Oil Tankers

Tankers are used to carry oil in bulk, and they can be divided into two different basic types. One type carries unrefined 'crude' oil while the other type, known as a product carrier, carries different types of 'refined' oils such as lubricating oil, naphtha, petrol or diesel oil.

The crude oil carriers are termed *very large crude carriers* (VLCC; 200 000–300 000 gt) or *ultra large crude carriers* (ULCC; 560 000+ gt). They are usually larger than the product carriers and can be as much as 300 000 gt. Students will appreciate the fact that if the crude oil is unrefined, then it will contain all the different types of oil all in one. This makes the crude oil quite volatile, and in the past there have been some significant explosions due to the mixture of hydrocarbons in a cargo tank.

Modern tankers are now fitted with fixed inert gas systems (all tankers over 8000 gt as of 1 January 2016). This means that when the oil is pumped from a tank during a discharge, the space above is filled with a gas that will not support a fire or explosion. Conversely as the oil is loaded the inert gas is released. More information on these systems can be found in Chapter 9 of this volume.

The machinery space and accommodation on oil tankers is situated aft. This means that the designers can provide an unbroken cargo space forward of these features. The cargo tanks are subdivided by longitudinal and transverse bulkheads, and the tanks are separated from the machinery space by an empty compartment known as a cofferdam. A pump room may also be provided at the after end of the cargo space just forward of the engine room and may form part of the cofferdam (Figure 1.3).

It is then possible to have the cargo pumps situated in the pump room and the prime mover (diesel engines, steam turbines or electric motors) situated in the machinery space. A gas tight seal is maintained around any rotating drive shaft penetrating the bulkhead between the machinery space and the pump room.

In the older vessels a double bottom was required only with respect to the machinery space and may have been used for the carriage of oil fuel and fresh water. Modern vessels must now have a 'double hull' covering the length of the ship. A forecastle is sometimes required and is used as a storage space, although on larger tankers this area is usually a continuation of the deck rather than a step change in the line of the ship. As the accommodation and navigation bridges are provided at the after end, the deck space may be left unbroken by superstructure and all the services and living arrangements, including catering equipment and facilities, are concentrated in one area.

In the smaller tankers much of the deck space is taken up by pipes and hatches. Therefore it is usual to provide a longitudinal platform or pathway to allow easy access to the *forecastle* and *bow* sections. The walkway is situated above the pipes and is known as the 'flying bridge'. On the VLCCs and ULCCs there is sufficient space to walk easily around the pipework. However the distance is so great that sometimes bicycles are provided for the crew. An alternative arrangement is for the ship to have a pump allocated to each cargo tank. Known as 'deepwell' pumps (see page 149) they have the prime mover sited on-deck and the pump at the bottom of each tank driven by a long drive shaft.

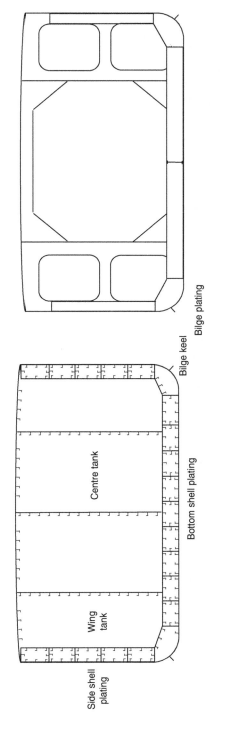

Double hulled oil tanker
general arrangement

Mid-section showing the web strengthening

Bilge keel

Bilge plating

Bottom shell plating

Centre tank

Wing
tank

Side shell
plating

Mid-section showing longitudinal arrangement

▲ **Figure 1.7** *Oil tanker – mid-section*

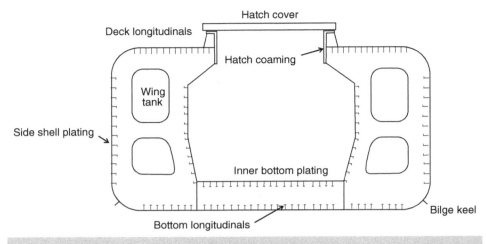

▲ **Figure 1.8** *Ore carrier – mid-section and general arrangements*

Another feature required by modern tankers is the ability to moor up to single buoy moorings (SBMs). These are typically arrangements where the output from an oil production field is fed along a pipe line, resting on the sea/river bed and leading to a mooring buoy that could be several miles away.

Readers can see that this single point, situated on the surface of the water, would warrant a special arrangement for securing the vessel. The Oil Companies International Marine Forum (OCIMF) produces guidelines for the design strength of systems for mooring to SBMs.

The method is to use the anchor chain; however, the angle of the pull on the securing equipment will have moved from the sea/river floor to the surface. Therefore, the weight on the equipment will be at a much shallower angle.

The midship section (Figure 1.7) shows the transverse arrangement of the cargo tanks. The centre tank is usually about half the width of the ship. However perhaps the most significant developments in tanker design is that of the inclusion of 'double hulls'. Following a series of mandates, during and just after the 1990s, regulation 19 in Annex 1 of MARPOL now requires that tankers over 5000 dwt be fitted with 'double hulls' or an alternative design approved by the IMO. More on this feature can be found in Chapter 9 of this volume.

The harmonised Common Structural Rules (CSR) for Bulk Carriers and Tankers were introduced on 1 July 2015. This set of rules, developed by the International Association of Classification Societies (IACS), replaced the rules set independently for bulk carriers and for double hull oil tankers. The rules are updated regularly in light of new information.

IACs are careful to explain that these rules cover self-propelled bulk carriers and double-hulled tankers, which are able to sail to any part of the world in almost any weather condition. The only restriction is sailing in icy conditions, which requires special arrangements.

The first and common part is arranged in 13 chapters and covers the minimum requirements for the strength of the hull, such as expected wave loads, hull girder strength as well as minimum buckling and fatigue characteristics. A design life of 25 years is assumed and forms the base for the size of the scantlings and so on.

The second part includes information specifically about the construction of the two different types of vessels.

Tankers five years old or more are subject to a 'special' survey. This survey will cover all the items in the 'annual' survey but will also examine all cargo tanks, ballast tanks, double bottom tanks, pump rooms, any pipe tunnels, cofferdams and void spaces. The aim is to ensure that the vessel is structurally sound for the next five years. The survey inspection will be backed up by hull thickness data and surveyors will look for corrosion, damage, fractures, deformation or set and any other structural deterioration.

The vessel will need to be dry-docked and all the crude oil washing and ballasting systems need to be in good working order. The tank coatings also need to be in good condition. The first special survey will require a spot check of the double hull but subsequent surveys will require a more rigorous inspection.

In addition to these inspections most respectable charters will insist on conducting a 'tanker vetting' process to ensure that the vessel is in a satisfactory condition to complete its charter and that it conforms to all the necessary requirements.

Tanker vetting will cover the condition of the vessel as well as inspect all the necessary records and documents to ensure that the tanker has been operated and maintained to the level required by the flag administration, by the classification society and by the insurers.

Bulk Carriers

Bulk carriers are vessels built to carry such cargoes as ore, coal, grain and sugar in large quantities. They are designed for ease of loading and discharging with the machinery space aft, allowing continuous, unbroken cargo space forward of the accommodation.

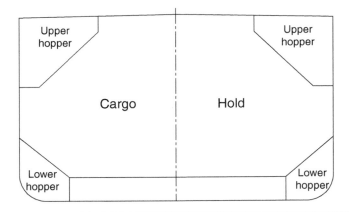

▲ **Figure 1.9** *Ore carrier*

They are single deck vessels having long, wide hatches, closed by steel covers. The double bottom runs from stem to stern.

In ships designed for heavy cargoes such as iron ore the double bottom is very deep and longitudinal bulkheads are fitted to restrict the cargo space (Figure 1.9). This system also raises the centre of gravity of the ore, resulting in a more comfortable ship. The double bottom and the wing compartments may be used as ballast tanks for the return voyage.

Some vessels, however, were designed to carry an alternative cargo of oil in these tanks. With lighter cargoes such as grain, the restriction of the cargo spaces is not necessary although deep hopper sides may be fitted to facilitate the discharge of cargo, either by suction or by grabs. The spaces at the sides of the hatches are plated in as shown in Figure 1.10 to give self-trimming properties. In the past many bulk carriers had a tunnel is fitted below the deck from the midship superstructure to the accommodation at the after end. The remainder of the wing space was used for water ballast. Some bulk carriers are built with alternate long and short compartments. Thus if a heavy cargo such as iron ore is carried, it is loaded into the short holds.

A cargo such as bauxite would be carried in the long holds, while a light cargo such as grain or timber would occupy the whole hold space.

The double bottom is continuous in the cargo space, and it is raised at the sides to form hopper sides which improve the rate of discharge of cargo. Wide hatches are fitted for ease of loading, while in some ships small wing tanks are fitted to give self-trimming properties.

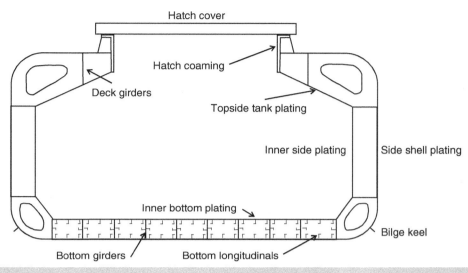

▲ **Figure 1.10** *Bulk carrier*

Chemical Carriers

A considerable variety of chemical cargoes are required to be carried in bulk. Many of these cargoes are highly corrosive and incompatible with each other while others require close control of temperature and pressure. Special chemical carriers have been designed and built, in which safety and avoidance of contamination are of prime importance.

To avoid corrosion of the structure, stainless steel is used extensively for the tanks, while in some cases coatings of zinc silicate or polyurethane are acceptable.

Protection for the tanks is provided by double bottom tanks and wing compartments which are usually about one-fifth of the midship beam from the ship side (Figure 1.11, lower drawing).

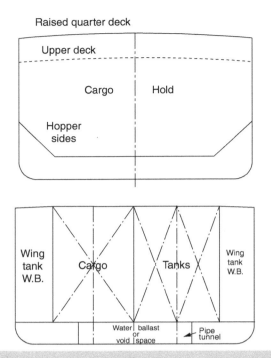

▲ **Figure 1.11** *Chemical carrier*

Liquefied Gas Carriers

Over the past 40 years the LNG and the liquefied petroleum gas (LPG) carriers have carved out their own class of vessel. The natural gas is mostly methane which may be liquefied by reducing the temperature to between –82°C and –162°C in association with pressures of 4.6 MN/m² to atmospheric pressure. The 'heavier' petroleum gas on the other hand consists of propane and butane and will become liquefied at a much higher temperature (–7°C).

Rules relating to the construction and operation of LNG carriers can be found in IMO International Code for the Construction and Equipment of Ships Carrying Liquefied Gases in Bulk (IGC Code). Since low carbon steel becomes extremely brittle at low temperatures, separate containers must be built within the hull and insulated from the hull. The tanks themselves are made from special steels such as nickel alloy steel or stainless steel. The most up-to-date systems have a double-walled membrane. Several different systems are available, one of which is shown in Figures 1.11 and 1.12.

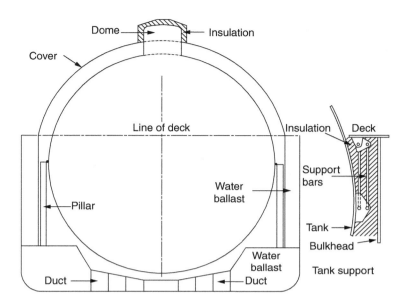

▲ **Figure 1.12** *Gas carrier with spherical tanks*

The cargo space consists of three large tanks set at about 1 m from the ship's side. Access is provided around the sides and ends of the tanks, allowing the internal structure to be inspected.

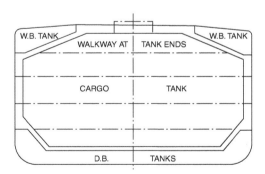

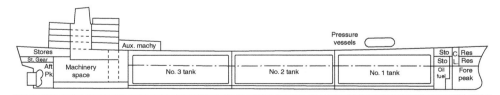

▲ **Figure 1.13** *Liquefied gas carrier*

Superyachts

This area of the industry is one that is expanding and developing fast. The demand for larger vessels has brought with it an increase in interest from regulators.

Until recently, superyachts were built to a fairly modest size and tonnage, and were operated by professional officers and crew on behalf of the owners. At that time the most important authority, IMO, did not have the time or resources to develop rules and guidelines for this area of the industry.

However during the end of the 1980s and through the 1990s the UK's MCA became increasingly concerned about the growing number and size of the vessels that make up this sector.

To offset the rising operating costs owners were also starting to turn to the practice of 'chartering' their yachts. This meant that there was a commercial undertone starting to take hold, and therefore the MCA felt the need to act without the blessing of the IMO. Twenty years on and flag administrations started to understand the wisdom of the MCA, but by then the third generation of the MCA's Large Yacht Code was being published. Version LY3 came about during 2015 and includes yachts carrying up to 12 passengers (Figure 1.14).

For yachts carrying 13–36 passengers, the Passenger Yacht Code (PYC) also applied. The second edition, introduced in 2012, sets out the requirements for yachts registered under the Red Ensign Group of flag administrations. In 2019 the LY3 was renamed the Red Ensign Yacht Code and comes in two sections, where part A is the Large Yacht Code and part B is the Passenger Yacht Code.

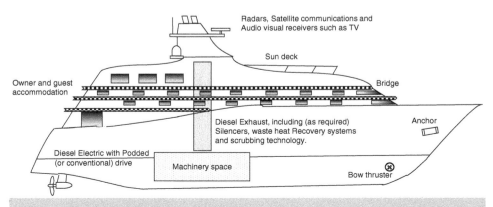

▲ **Figure 1.14** *Superyacht – general arrangement*

The rules cover information about the:

- strength and standard of design of a yacht's structure
- water and weathertight integrity
- requirements for the machinery and electrical installation
- steering arrangements
- bilge pumping requirements
- stability and freeboard
- lifesaving equipment required
- fire resistance and firefighting equipment
- radio communication
- navigation equipment

As well as things such as the use of helicopters and tenders/survival craft that are specific to yachts. The regulations also cover the requirements for the qualifications of the officers and crew that operate the yachts.

General Notes

It used to be that ocean-going ships were able to exist as totally independent units. The cargo handling equipment suitable for the ship's service was carried with the vessel. That aspect has now changed for some ships, notably container ships and large bulk carriers. Complex and sophisticated electronic navigation aids, radar and global satellite positioning and communication equipment have now mostly replaced paper charts and radio equipment.

The main and auxiliary machinery must be sufficient to propel the ship at the required speed and to maintain the ship's services efficiently and economically. Adequate redundancy and backup for all essential services are required. Accommodation, at the required standard (MLC 2006), must be provided for officers and crew with comfortable cabins, recreation rooms and dining rooms. Many ships are now also being fitted with wired or wireless internet access.

Efficient heating, ventilation and air conditioning (HVAC) systems are fitted due to the tremendous variation in air temperature when travelling to different parts of the globe. Bridges and bridge wings are now totally enclosed, and there is an air conditioned area in the machinery space called the machinery control room.

Manoeuvrability is greatly improved with the inclusion of bow and stern thrusters and/ or controllable pitch propellers. The recent development of 'podded' drives has also improved manoeuvrability a stage further.

Many ships have small swimming pools and the ships must carry sufficient foodstuffs in refrigerated and non-refrigerated stores to last at least until the next port where stores are available. Drinking water is obviously essential. Desalination or reverse osmosis plants can usually recover enough fresh water from sea water to supply the needs of the crew and machinery. In the event of emergency it is essential that first aid, fire extinguishing and life-saving appliances are provided.

IMO now require that ships are constructed to a 'risk' assessed standards called goal based standards for bulk carriers and tankers SOLAS Ch II-1/2.28 which brings them into line with passenger ships (Figures 1.14 and 1.15).

However, the main focus currently, is about reducing the environmental footprint of the maritime industry. Efficiency gains are very important, as these reduce a ship's carbon footprint and also reduce the cost of running the ship. Efforts are centered on reducing air pollution, but other areas of concern are ballast water (transportation of invasive species) and hull coatings.

Ship Terms

The following terms and abbreviations are in use throughout the shipbuilding industry.

Length overall (LOA)
The distance from the extreme fore part of the ship to a similar point aft and is the greatest length of the ship. This length is important when docking.

Length between perpendiculars (LBP)
The forward perpendicular is the point at which the summer load waterline crosses the stem. The aft perpendicular is the after side of the rudder post or the centre of the rudder stock if there is no rudder post. The distance between these two points is known as the length between perpendiculars, and is used for some ship calculations.

Breadth
The greatest breadth of the ship, measured to the outside of the shell plating.

Breadth moulded (BMld)
The greatest breadth of the ship, measured to the inside of the inside strakes of shell plating.

Bulwark
Vertical plating that extends upwards and is fitted around the perimeter of the main deck or weather deck.

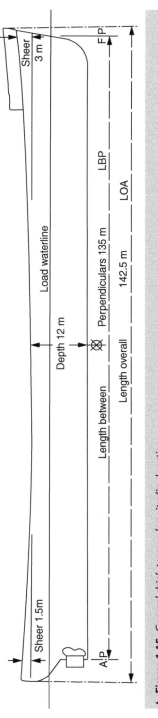

▲ **Figure 1.15** *General ship's terms – longitudinal section*

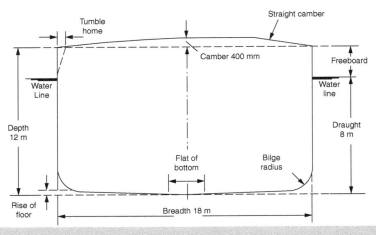

▲ **Figure 1.16** *General ship's terms – transverse section*

Coaming

Vertical side of the hatch extending from the main deck and forming a structure for the hatch lid to sit upon.

Cofferdams

Void space that sits between two different spaces or tanks that are usually carrying different liquids. The cofferdam than acts as a gap between the tanks, preventing contamination in the event of a leakage from one of the tanks.

Collision bulkhead

Name given to the watertight bulkhead situated the furthest forward and is therefore the first bulkhead that will be required to take the force of an impact at the forward end of the vessel.

Depth extreme (DExt)

The depth of the ship measured from the underside of the keel to the top of the deck beam at the side of the uppermost continuous deck amidships.

Depth moulded (DMld)

The depth measured from the top of the keel.

Double bottom

Name given to the area that includes the outer hull, girders and stiffeners inside the vessel and a layer of plating to form in effect a double skin. This should not be mixed up with the new form of 'twin hulled' tankers. This is due to the double bottom being used as a space for liquid storage and the twin hull arrangement being an empty space.

Draught extreme (dExt)

The distance from the bottom of the keel to the waterline. The load draught is the maximum draught to which a vessel may be loaded. This will vary depending upon the service and type of water. (See Chapter 10 for more details.)

Draught moulded (dMld)
The draught measured from the top of the keel to the waterline.

Freeboard
The distance from the waterline to the top of the deck plating at the side of the deck amidships.

Camber or round of beam
The transverse curvature of the deck from the centreline down to the sides. This camber is used on exposed decks to drive water to the sides of the ship. Other decks are often cambered. Most modern ships have decks which are flat transversely over the width of the hatch or centre tanks and slope down towards the side of the ship.

Sheer
The curvature of the deck in a fore and aft direction, rising from midships to a maximum at the ends. The sheer forward is usually twice that aft. Sheer on exposed decks makes a ship more seaworthy by raising the deck at the fore and after ends further from the water and by reducing the volume of water coming on the deck.

Rise of floor
The bottom shell of a ship is sometimes sloped up from the keel to the bilge to facilitate drainage. This rise of floor is small, 150 mm being usual.

Bilge radius
The radius of the arc connecting the side of the ship to the bottom at the midship portion of the ship.

Bilge keel
A section of plating fixed to the outside of the hull running for the length of the ship protruding at right angles to the bilge radius.

Tumble home
In some ships the midship side shell in the region of the upper deck is curved slightly towards the centre line, thus reducing the width of the upper deck and decks above. Such tumble home improves the appearance of the ship.

Displacement
This is a measurement of the mass of the ship and everything it contains when the measurement is taken. The term comes from the amount of water that a ship will 'displace' when it is fully floating. Please note that there will be a difference between a ship floating in fresh water and the same ship, loaded exactly the same, floating in salt water. This is due to the difference in density between the salt water and fresh water. Displacement can be calculated as the underwater volume times by the density of the water that the vessel is floating in, times by the value of gravity.

Lightweight

This is a measure of the mass of the empty ship, without stores, fuel, water, crew or their effects. The hull and machinery and all the fixtures and fittings are also included in this measurement.

Deadweight

The deadweight is a measure of the mass that a ship is carrying at a given time. It is the sum of the weight of cargo, fuel, water, stores and people that a ship has on board when the measurement is taken. The deadweight is therefore the difference between the displacement and the lightweight:

Displacement = Lightweight + Deadweight

It is usual to categorise a vessel by reference to its deadweight. Thus a 10 000 tonne ship is one which is capable of carrying a deadweight of 10 000 tonne.

Registered tonnage

It is necessary to have an official measurement for ships and in the past the value of the gross registered tonnes (grt) has been used. However there was never a universally agreed standard definition of grt.

IMO's International Convention on Tonnage Measurement of Ships entered into force in July 1982 and as a consequence the two measurements of gross tonnage (gt) and net tonnage (nt) have been agreed upon and are now in universal use for all ships.

However, they are not straightforward mathematical calculations. IMO describe 'gross tonnage' as a function of the volume of all the internal spaces within the ship. These include the volume of appendages but not volumes that are open to the sea.

The volumes are not simply added up and the actual calculation is:

Gross tonnage = K1 V,

where K1 = 0.2 + 0.02 log10 × V and V = the total volume of all the qualifying spaces of the ship in cubic metres.

Net tonnage is calculated by using a mathematical function of the volume of all the cargo spaces of the ship but it cannot be recorded as being more than 30% below the gross tonnage.

Uppermost continuous deck or bulkhead deck

This is one of the most important features of a ship as it makes a watertight seal with the vertical watertight bulkheads. With cargo ships, this deck could also be the same as the freeboard deck.

This deck should be provided with the means to close all openings that could be accessed by the sea, thus making a sealed watertight box with the watertight bulkheads.

The uppermost continuous deck or bulkhead deck is also taken as the 'strength' deck when calculating the girder strength of the vessel.

2

STRESSES IN A SHIP'S STRUCTURE

A number of forces act on a ship's structure; some are static forces while others are dynamic. The static forces are set up due to the differences in weight and support along the length of the ship, while the dynamic forces are created by the force of the wind pushing on the ship, as well as the water interacting with the ship, by the passage of waves along the ship and by the moving propulsion parts. The greatest stresses set up in the ship as a whole are due to the distribution of loads along the ship, causing longitudinal bending.

Strength of Ships – Overall Design Concept

Students will be able to see that this aspect is the fundamental starting point for calculating the overall strength of a ship. As would be expected, computers are now used extensively to assist the ship's designer to calculate the ultimate hull girder strength.

A ship may be regarded as a non-uniform beam, carrying both uniformly distributed and non-uniformly distributed loads and having varying degrees of support along its length. Some of the load will be distributed evenly over a section of the ship while some will be more concentrated. The overall strength of the beam is referred to as the 'girder strength' and the overall bending moment envelope curves are used to calculate the required girder strength for all circumstances that the vessel is likely to encounter. Then the internal structures and the hull plating are sized and arranged to meet the minimum girder strength required for the design and duty of the vessel.

Values for the still water bending moments, under all conditions of operation, are calculated. The results of all the bending moments give the overall bending moment for the hull girder, and this divided by the maximum design stress allowed by the classification society, will give the strength modulus required for the hull girder steel sections. (See Chapter 9 of *Reeds Vol 2* for more information about the calculation of stress and modulus of sections.)

The components designed to resist the buckling of the girder and contributing to the ultimate hull strength are the:

- size, number and strength of longitudinal beams
- thickness and strength of the shell plating
- bilge keels
- quality of welding
- number and strength of the transverse sections.

The ultimate hull girder strength for tankers, bulk carriers and now container ships is determined using a system known as the iterative-incremental method, which is where the hull girder is divided into a set of transverse elements. The forces are calculated in one element and the resultant strain (for any one given condition) is used to modify the stress calculations on the next connecting transverse element. This is repeated until the final strength profile for the hull girder has been calculated.

If we now think about the hull girder, students will be able to see that to keep the 'box section' in shape there needs to be a combination of beams running along the length of the box and a series of shapes designed to maintain the square shape of the box (see Figure 2.8). The way that these two are combined will determine the overall design of the vessel.

There are two different systems of framing, called longitudinal framing and transverse framing. As the two names suggest, longitudinal framing is where the main load carrying sections of steel run parallel to the sides of the vessel. Transverse framing on the other hand, has the framing arranged at right angles to the sides of the ship. These frames are then supported by girders, or stringers, running in parallel with the sides of the vessel. Not only can the overall vessel be of one or the other or indeed a combination of both, but so can smaller parts of the overall vessel, with respect to hatch openings for example.

Special consideration must be given to any transition from one system to the other. Any transition must be carefully considered by the designer and by the quality assurance processes of the construction yard.

Longitudinal Bending

Still water bending – static loading

If we consider a loaded ship lying in still water, then the upthrust at any 1 m length of the ship depends upon the immersed cross-sectional area of the ship at that point. If the values of upthrust at different positions along the length of the ship are plotted on a base representing the ship's length, a *buoyancy curve* is formed (Figure 2.1).

This curve increases from zero to a maximum value in the midship portion, then decreases back down to zero. The area of this curve represents the total upthrust exerted by the water on the ship. The total weight of a ship consists of a number of independent weights concentrated over short lengths of the ship. These include; cargo, machinery, accommodation, cargo handling gear, poop and forecastle sections

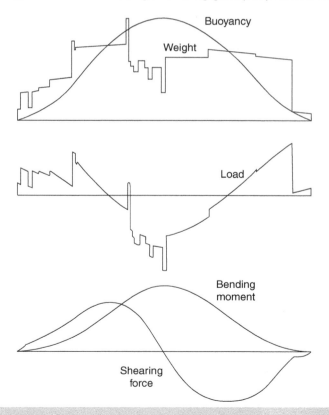

▲ **Figure 2.1** *Load distribution*

of the hull construction, and a number of items which form continuous material over the length of the ship, such as decks, shell and tank top.

A *curve of weights* is shown in Figure 2.1. The difference between the weight and buoyancy at any point is the load at that point. In some cases the load is in excess of weight over buoyancy and in other cases there is an excess of buoyancy over weight. A load diagram formed by these differences is shown in the figure. Since the total weight must be equal to the total buoyancy (assuming that the vessel is still floating), the area of the load diagram above the base line must be equal to the area below the base line.

Due to the unequal loading, however, shearing forces and bending moments are set up in the ship with the maximum bending moment occurring around the midship section.

The load distribution will determine the direction in which the bending moment will act, and this in turn will create the state of hogging or sagging. Class nomenclature for the condition of hogging and sagging in the bending moment calculations is to go negative for sagging and positive for hogging.

If, for example, the buoyancy amidships exceeds the weight, the ship will hog, and this may be likened to a beam supported at the centre and loaded at the ends.

As with a simply supported beam, when a ship hogs, the deck structure is in tension while the bottom plating is in compression (Figure 2.2). If the weight amidships exceeds the buoyancy, the ship will sag, which is equivalent to a beam supported at its ends and loaded at the centre.

When a ship sags, the bottom shell is in tension while the deck is in compression (Figure 2.3). Students will be able to appreciate that when a hull is continuously changing between hogging and sagging, as in a rough sea, considerable cyclical stresses happen in the deck and the bottom shell plating.

Changes in bending moments also occur in a ship due to different loading conditions. This is particularly true in the case of cargoes such as iron ore which are heavy compared

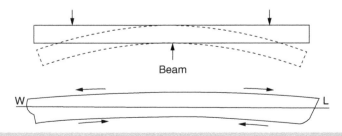

Beam

W L

▲ **Figure 2.2** *Hogging*

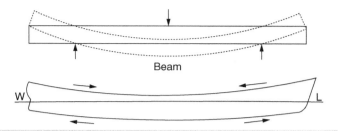

▲ **Figure 2.3** *Sagging*

with the volume they occupy. When these types of cargo are loaded into a ship, especially if it is on the spot market or performing the role of a tramp ship, care must be taken to ensure a suitable distribution throughout the ship. The even distribution of stresses is calculated by using the on-board loading computer.

In the past these calculations have proved difficult especially if the ship has a machinery space and deep tanks/cargo holds amidships. These older vessels would also have had only a very basic method of calculating the bending moment. There would however be a tendency in such ships, when loading heavy cargoes, to leave the deep tank empty. This results in an excess of buoyancy by way of the deep tank. This action must be considered carefully as there could also be an excess of buoyancy by way of the engine room, since the machinery (especially if large two stroke engines are fitted) might be light when compared with the volume it occupies. The International Association of Classification Societies (IACS) sets out guidance for the loading sequences in its rules for the 'Strength of Ships'. Careful consideration should also be given to the sequencing of loading and using bunker fuel as well as the filling or emptying of the ballast tanks.

A ship loaded carelessly, might hog considerably, creating unusually high stresses in the deck and bottom shell. This may be very dangerous and could lead to the vessel breaking in two if loaded using an incorrect sequence. If the owners intend for the ships to be regularly loaded in this manner, additional hull strength must be provided to ensure the safe operation of the vessel. In cases where there is a long transmission shaft between the main engine and the propeller, excess hogging or sagging could also lead to excessive bending of this shaft and the engineering staff would continually be checking for any overheating of the main shaft bearings.

Highly sophisticated computer tracking systems monitor the movement of containers as they are transported around the world. Therefore, container ships will have their loading and discharging sequencing calculated before the vessel reaches port. This means that the stresses and bending moments will also be calculated while the vessel is moving between ports.

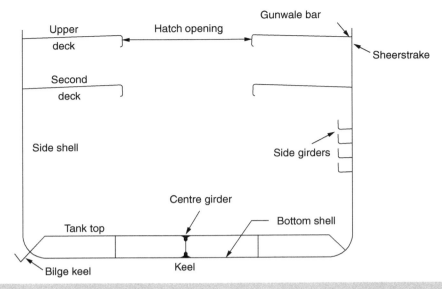

▲ **Figure 2.4** *Longitudinal material*

The structure resisting longitudinal bending is made up of all the continuous longitudinal material. The features farthest from the axis of bending (the neutral axis) are the most important (Figure 2.4). These features are the:

- keel
- bottom shell plating
- centre girder
- side girders
- tank top
- tank margin
- side shell
- sheerstrake
- stringer plate
- deck plating alongside hatches
- and in the case of oil tankers, any longitudinal bulkheads.

Buckling and/or deformation may occur at a point in the structure that is the greatest distance from the neutral axis which will become a high stress point, such as the top of a sheerstrake. Designers work to ensure that such points are avoided as far as possible, since they may result in the plate cracking. In many oil tankers the structure is improved by joining the sheerstrake and stringer plate to form a rounded gunwale.

Wave bending – dynamic loading

When a ship passes through waves, alterations in the distribution of buoyancy cause alterations in the bending moment. The greatest differences occur when a ship passes through waves whose lengths from crest to crest are equal to the length of the ship thereby placing the greatest bending moment on the hull.

When the wave crest is amidships (Figure 2.5), the buoyancy amidships is increased while at the ends it is reduced. This tends to cause the ship to hog. A few seconds later the wave trough lies amidships. The buoyancy amidships is reduced while at the ends it is increased, causing the vessel to sag (Figure 2.6).

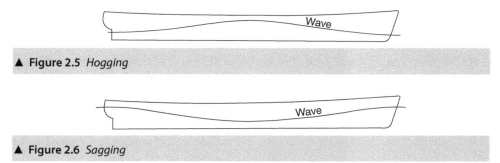

▲ **Figure 2.5** *Hogging*

▲ **Figure 2.6** *Sagging*

The effect of these waves is to cause fluctuations in stress, or, in extreme cases, complete reversals of stress every few seconds. The ship is designed to withstand this cycle of stressing without causing undue damage. However, if any part of the structure has already been damaged or is corroded, then the hogging and sagging will make the weakened structure worse.

Transverse Bending and Racking

Consider the cross-section of a vessel at a point along its length (see Figure 2.7). The transverse structure may be subjected to three different types of loading. These are forces:

- from the weight of the ship's structure, machinery, fuel, water and cargo
- due to the water pressure
- causing longitudinal bending.

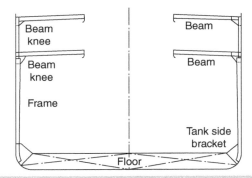

▲ **Figure 2.7** *Transverse material*

The decks must be designed to support the weight of accommodation, winches and cargo, while exposed decks may also have to withstand a tremendous weight of water that might be taken on board during heavy weather. The deck plating is connected to beams which transmit the loads to the longitudinal girders and to the side frames.

In the area of heavy local loads such as cranes and windlasses and so on, additional stiffening will be required. The shell plating and frames form pillars which support the additional weights that are situated on the deck. Tank tops are required to be strong enough to keep the cargo in place or resist the upthrust exerted by the liquid in the tanks.

In the machinery space other factors must be taken into account. Fluctuating forces transmitted from the reciprocating machinery through to the structure need to be accommodated. Modern resilient mountings reduce the magnitude of the forces and the strength of the fixings means that the machinery is well supported to prevent any excess movement.

Under the position of the engine additional girders are fitted in the double bottom and the thickness of the tank top increased to ensure that the main propulsion remains fixed despite the additional stresses caused by rough weather acting upon the vessel.

Special consideration must be given to the thrust block, the propeller shaft and the propeller. Thrust to push the ship along is generated by the propeller and must be carefully transmitted to the hull of the vessel. This is a difficult process as the propeller shaft is relatively small in diameter when compared with the area of the hull.

The thrust block is first in line to take the force from the propeller shaft. The important issue is for the thrust block to be connected to as large an area of the hull as possible. This will transmit the force generated by the thrust to the hull evenly.

The weight of the propeller shaft is supported by intermediate bearings which in turn must be supported by the vessel's hull structure. Accommodating the weight of the propeller is particularly interesting as this revolving mass is on the end of the tail shaft which is supported by the stern tube bearing. The arrangement of the stern of the vessel needs to counter the forces transmitted through the stern tube bearing. These forces will mostly be the weight of the propeller which will be acting at the end of the shaft. Any out of balance forces will have a significant effect as will any vibration caused by cavitation. These are important reasons for ensuring that the propeller remains undamaged.

A considerable force is exerted on the bottom and side shell by the water surrounding the ship and the double bottom floors and side frames are designed to withstand these forces, while the shell plating must be thick enough to prevent buckling as it spans the distance between the floors and frames.

Since water pressure increases with the depth of immersion, the load on the bottom shell plating will be greater than the load on the side shell. It follows, therefore, that the bottom shell must be thicker than the side shell to withstand the increased force. When the ship passes through waves, these forces are of a pulsating nature and may vary considerably in high waves, while in bad weather conditions the shell plating above the waterline will receive severe hammering.

When a ship rolls there is a tendency for the ship to distort transversely due to the fluctuating forces described above and in a similar way to that shown in Figure 2.8. This action is known as *racking* and is reduced or prevented by the beam knee and tank side bracket connections, together with the transverse bulkheads, the latter having the greatest effect.

The efficiency of the ship structure in withstanding longitudinal bending depends to a large extent on its girder strength and the ability of the transverse structure to prevent the buckling of the shell plating and decks.

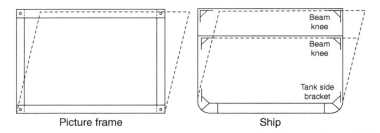

Picture frame　　　　　　　　　Ship

▲ **Figure 2.8** *Racking*

Dry-docking

A ship usually enters dry dock with a slight trim aft. This means that as the water is pumped out, the after end touches the blocks first. As more water is pumped out an upthrust is exerted by the blocks on the after end, causing the ship to change trim until the whole keel from forward to aft rests on the centre blocks. At the instant before this occurs the upthrust aft is at maximum. If the design of the ship results in this thrust being excessive, it may be necessary to strengthen the after blocks and the after end of the ship. Such a problem arises if it is necessary to dock a ship when fully loaded or when trimming severely by the stern.

As the pumping continues the load on the keel blocks is increased until the whole weight of the ship is taken by the blocks in the dry dock. The ship structure must be strong enough to withstand this unevenly distributed load. The 'docking' plan, carefully worked out before the ship arrives, ensures that the blocks are all placed in the correct position.

The strength of the hull is carefully considered during the design of the ship and there could be up to three different docking arrangements specified for each vessel. Different systems can be used on subsequent dry docks so that the spaces not examined at the last docking can be covered during the current one. It will also be obvious to students that the hull cannot be prepared and painted in the area in contact with the blocks.

In most ships the normal arrangement of keel and centre girder, together with the transverse floors, is quite sufficient for the purpose. If a duct keel is fitted, however, care must be taken to ensure that the width of the duct does not exceed the width of the keel blocks (Figure 2.9). The keel structure of a longitudinally stressed vessel

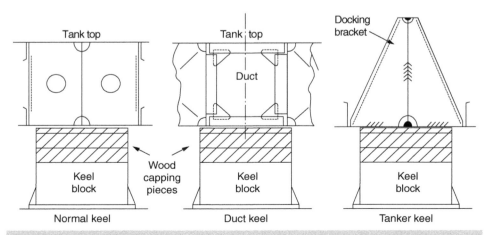

▲ **Figure 2.9** *Examples of bottom arrangements for docking*

such as an oil tanker, bulker or container ship is strengthened by fitting docking brackets and tying the centre girder to the adjacent longitudinal frames at intervals of about 1.5 m.

Bilge blocks or shores could be fitted to support the sides of the ship. The arrangements of the bilge blocks vary from dock to dock. In some cases they are fitted after the water is pumped out of the dock, while other dry docks may have blocks which can be slid into place while the water is still in the dock. The latter arrangement is preferable since the sides are completely supported. At the ends of the ship, where the curvature of the shell does not permit blocks to be fitted, bilge shores are used. The structure at the bilge must prevent these shores and blocks buckling the shell.

As soon as the after end touches the blocks, shores are inserted between the stern and the dock side to centralise the ship in the dock and to prevent the ship slipping off the blocks. When the ship grounds along its whole length additional shores could be fitted on both sides, holding the ship in position and preventing tipping. These shores are known as *breast shores* and have some slight effect in preventing the side shell bulging. They should preferably be placed with respect to transverse bulkheads or side frames as these offer more resistance to buckling than the side placed do on their own (Figure 2.10). On the larger vessels the side supports are not necessary as the ship sits safely on the blocks situated on the floor of the dock.

When undocking the vessel, care must be taken to ensure that all the ballast, fresh water and fuel oil tanks are in the same condition as they were when the vessel came into the dry dock.

It is usual to start filling the dock and then, when the ship's side valves are just covered, stop the filling to check for any leaks.

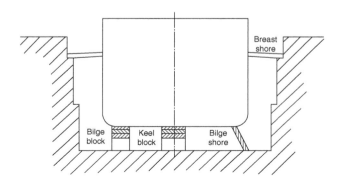

▲ **Figure 2.10** *Support in dock*

Pounding

When a ship meets heavy weather and commences heaving and pitching, the rise of the fore end of the ship occasionally synchronises with the trough of a wave. The fore end then emerges from the water and re-enters with a tremendous slamming effect, known as pounding. While the event does not occur with great regularity, it may nevertheless cause damage to the bottom of the ship at the forward end. The designers must ensure that the shell plating is stiffened to prevent buckling due to the forces involved in the pounding. Pounding also affects the aft end section of the vessel but the effects are not nearly as great. Nevertheless, provision must be made in the design of the hull to counteract the effects of pounding at the aft end.

Panting

As waves pass along the length of a ship the various parts of the vessel are subjected to varying depths of water which causes fluctuations in water pressure. This tends to create an in-and-out movement of the shell plating. The knock-on effect of this is found to be greatest at the ends of the ship, particularly at the fore end, where the shell is relatively flat. Such movements are termed panting, and, if unrestricted, panting could eventually lead to fatigue of the material and must therefore be prevented as much as possible. This is achieved by the structure at the ends of the ship being stiffened to prevent any undue movement of the shell plating (see Chapter 7 for more details).

3

STEEL SECTIONS USED: WELDING AND MATERIALS

Introduction

When iron was used in the construction of ships in preference to wood, it was found necessary to produce forms of the material suitable for connecting plates together and acting as stiffeners. These forms were called sections and were produced by passing the material through suitably shaped rollers. The development of these bars continued with the introduction of steel until many different sections were produced. These sections are used in the building of modern ships and are known as *rolled steel sections*. The 'rolling' of the steel means that the internal 'grain' is aligned in the same direction adding to the strength of the section. The most common forms are described over the next few pages.

Ordinary angles

These sections may be used to join together two plates meeting at right angles or to form light stiffeners in riveted ships. Two types are employed: those having equal flanges (Figure 3.1), varying in size between 75 mm and 175 mm, and those having unequal flanges (Figure 3.2), which may be obtained in a number of sizes up to 250 mm by 100 mm, the latter type being used primarily as stiffeners.

▲ **Figure 3.1** *Equal angle*

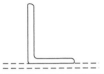

▲ **Figure 3.2** *Unequal angle*

▲ **Figure 3.3** *Toe welded angle*

▲ **Figure 3.4** *Bulb plate*

In welded ships, connecting angles are no longer required but use may be made of the unequal angles by toe-welding them to the plates, forming much more efficient stiffeners (Figure 3.3).

Bulb plates

A bulb plate (Figure 3.4), having a bulb slightly heavier than the older and now unused bulb angle, has been specially developed for welded construction. A plate having a bulb on both sides has been available for many years, but its use has been severely limited due to the difficulty of attaching brackets to the web with respect to the bulb. The modern section resolves this problem since the brackets may be either overlapped or butt welded to the flat portion of the bulb. Such sections are available in depths varying between 80 mm and 430 mm, being lighter than the bulb angles for equal strength. They are used for general stiffening purposes in the same way as bulb angles.

Channels

Channel bars (Figure 3.5) are supplied in depths varying between 160 mm and 400 mm. Channels are used for panting beams, struts, pillars, girders and heavy frames. In insulated ships it is necessary to provide the required strength of bulkheads, decks and shell with a minimum depth of stiffener and at the same time provide a flat inner surface for connecting the facing material in order to reduce the depth of insulation required and to provide maximum cargo space. In many cases, therefore, channel bars with reverse bars are used for such stiffening (Figure 3.6), reducing the depth of the members by 50 mm or 75 mm. Both the weight and the cost of this method of construction are high.

▲ **Figure 3.5** *Channel bar*

▲ **Figure 3.6** *Channel bar and reverse*

Joist or H-bars

These sections have been used for many years for such items as crane rails but have relatively small flanges. The manufacturers have now produced such sections with wide flanges (Figure 3.7), which prove much more useful in ship construction. They are used for crane rails, struts and pillars, being relatively strong in all directions. In deep tanks and engine rooms where tubular pillars are of little practical use, the broad flanged beam may be used to advantage.

T-bars

The use of the T-bar (Figure 3.8) is limited in modern ships. Occasionally they are toe-welded to bulkheads (Figure 3.9) to form heavy stiffening of small depth. Many ships have bilge keels incorporating T-bars in the connection to the shell.

▲ **Figure 3.7** *Broad flanged beam*

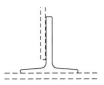

▲ **Figure 3.8** *T-bar*

Weld

▲ **Figure 3.9** *T-bar toe welded*

Flat bars or slabs

Flat bars are often used in ships of welded construction, particularly for light stiffening, waterways and save-alls which prevent the spread of oil. Large flat bars are used in oil tankers and bulk carriers for longitudinal stiffening where the material tends to be in tension or compression rather than subject to high bending moments. This allows for greater continuity in the vicinity of watertight or oil-tight bulkheads.

Several other sections are used in ships for various reasons. Solid round bars (Figure 3.10) are used for light pillars, particularly in accommodation spaces, for welded stems and for fabricated rudders and stern frames. Half-round bars (Figure 3.11) are used for stiffening in accommodation where projections may prove dangerous (e.g. in toilets and wash places) and for protection of ropes from chafing.

▲ **Figure 3.10** *Solid round*

▲ **Figure 3.11** *Half-round*

Aluminium sections

Aluminium alloys that are used in ship construction have been found to be too soft to roll successfully in section form and are therefore produced by extrusion. This method produces the different shapes by forcing the metal through a suitably shaped die. This becomes an advantage since the dies are relatively cheap to produce, allowing numerous different shapes of section to be made.

There are a few *standard sections* but the aluminium manufacturers are prepared to extrude any feasible forms of section which the shipbuilders require in reasonable quantities. Figure 3.12 shows examples of some such sections, which have been produced for use on ships built in this country.

(a)

(b)

▲ **Figure 3.12** *Aluminium sections*

Welding

Welding is the process of using heat to melt two separate pieces of metal and joining them together in such a way that they become one integral unit. It uses an energy source to produce the heat, such as compressed gas and/or electric currents. There are two basic types of welding: resistance or pressure welding, in which the portions of metal are brought to a welding temperature and an applied force is used to form the joint; and fusion welding, where the two parts forming the joint are raised to a melting temperature and either drawn together or joined by means of a filler wire of the same material as the adjacent members. The application of welding to shipbuilding is almost entirely restricted to fusion welding in the form of metallic arc welding. However, on board the ship, gas welding and cutting is also a very important and useful maintenance and repair tool.

Welding Safety

The safety of the staff on board modern ships is obviously very important and more so as some of the activities can be potentially hazardous. Welding is one of those activities that can go very wrong and cause major problems if precautions are not taken by trained and knowledgeable staff.

Any proposed work must be discussed at the safety brief and the necessary 'permit to work' must also be issued. The ship's safety management system is also available to guide staff on the specific company procedures necessary. Some ship's may also carry the UK's 'Code of Safe Working Practices for Merchant Seafarers', where Chapter 24 contains detailed advice for sea staff about the dangers associated with welding.

General

The most obvious hazard, which is common across the different welding techniques, is due to heat and sparks. Students will be able to see that these have the ability to start a fire if action is not taken to guard against them.

It is fairly obvious that all combustible material must be removed from the area surrounding the 'hot work' and fire extinguishing equipment kept ready nearby. However, it is less obvious to check the other side of metal bulkheads and deck heads especially if welding is being undertaken close to one of these features. Fires have also been started by sparks falling into the bilges where oil could be present.

People also become complacent about wearing protective clothing; however, when welding, at all times, protective clothing is important. In this case protection is needed against the heat causing burns, which means protective gloves and possibly additional heat resistant clothing over ordinary work clothes. Safety boots should be worn at all times when working on board, however, it is important to check that they are in good condition and suitable for 'welding'.

The welding process gives off fumes, and the composition of the fumes depends upon the method of welding taking place, the filler rods being used and the types of metal being welded. Nitrous oxide, for example, is one of the main gases present in welding fumes.

The important rule for any welding is to complete the process in a well ventilated area. If this is not possible, such as inside a small tank, then the welder must be provided with suitable fresh air breathing equipment.

The welding equipment keeps oxygen away from the actual process as much as possible, by surrounding the melting metal with a form of protective coating. When the welded metals are cooling the protection forms a hard, brittle outer shell. When this shell, called slag, is removed using a chipping hammer, there is a possible danger of sharp pieces flying towards the eyes and causing harm if the eyes are unprotected. The eye protection for this task involves using 'clear' goggles as opposed to the darkened glasses that are used for the actual welding. Changing from the darkened glass to the clear glass goggles is often overlooked, and removing slag is often then completed with unprotected eyes. The most diligent staff won't be caught out by not using the correct eye protection.

Manual metal arc welding (electric arc welding)

The ultra-violet light given off by the manual metal arc (MMA) welding process is much more intense than with the gas welding process, and therefore it is vital to use the correct standard of darkened glass in the welding mask. It is also very important to check that the eye protection is in good condition, with no damage, and carries the correct international quality standard and its use is understood by the person undertaking the welding. It's important to remember that not only is the welder vulnerable to eye damage from arc welding, but people passing by are also susceptible if they happen to look at the electrode just as the welder strikes an arc.

Electric shock is also a danger when using arc welding equipment. It is important that the equipment and surrounding areas are kept dry.

The wires connections and equipment must be checked for damage before any welding is undertaken. Faulty equipment could lead to overheating, electric shock and/or further damage to the equipment. It is also best not to trail the wires through any water that could be laying on the metal decks.

Gas welding – eyes

As with electric arc welding, eye protection is extremely important in both gas welding and cutting. The problem is that there is always the temptation to discard the goggles when using the gas equipment, thinking that the light given off from the gas welding process can be viewed with the naked eye. This is *not* the case and eyesight damage is bound to happen with prolonged exposure to the gas welding process.

Careful attention must be paid to assembling the equipment. There are *two* separate systems that are kept apart until the final flame at the end of the torch. A combustible gas, usually acetylene, is used in one system and oxygen in the other. The two systems are *both* colour coded and given incompatible fittings, so that an acetylene hose cannot be used on an oxygen fitting. Hoses, connections and equipment must be checked for damage before use. The thread on the connections are arranged to couple up in different directions. The oxygen has a conventional 'right'-handed system while the combustible gas has the less common 'left'-handed thread system. Care must be taken not to force the nut from one system onto the fitting of another. If by some outside chance and by using considerable force this was accomplished, the threads will be stripped off and leakage will occur.

Welding Processes

Manual metal arc welding

This is sometimes known as 'stick' welding or 'arc' welding. Figure 3.13 shows a simplified circuit that is used in arc welding. A metal electrode, of the same material as the work piece, is clamped into a holder which is connected to one terminal of a welding unit, the opposing terminal being connected to the work piece. An arc is formed between the electrode and the work piece or metal to be welded close to the position of the joint. The arc between the two metals to be joined, completing the circuit, creates a very high temperature which melts the two parts of the joint and the electrode. Metal particles from the electrode then transfer to the work piece, forming the weld.

The arc and the molten metal must be protected from the atmosphere to prevent oxidation. When steel is being welded a coated electrode is used, the coating being in the form of a silicone. This coating melts at a slightly slower rate than the metal and is carried with the metal particles to form a slag over the molten metal, while at the same

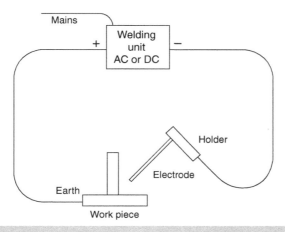

▲ **Figure 3.13** *Welding circuit*

time an inert gas is formed which creates a shield over the arc (Figure 3.14). When the weld and the parent metal have cooled, the slag is removed from over the weld by chipping with a special hammer. Care must be taken as the slag is brittle and can fly off in unexpected directions.

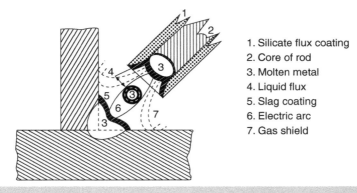

1. Silicate flux coating
2. Core of rod
3. Molten metal
4. Liquid flux
5. Slag coating
6. Electric arc
7. Gas shield

▲ **Figure 3.14** *Weld arc*

The MMA equipment can be one of three different types. The first, and least used, type is where the equipment uses alternating current (AC) to the machine and through the welding circuit. This design could be particularly hazardous to operators, resulting in possible heart failure in some instances. AC as the welding force is helpful when using tungsten inert gas (TIG) welding equipment and, therefore, the MMA system is much better suited to using one of the following types of direct current (DC) arrangement for the welding circuit.

The DC circuit can be connected with the electrode on the positive or the negative side (Figure 3.13). Each method produces quite a different result and, therefore, seafarers must be very careful about connecting the electrode to the correct polarity. Connecting the rod to the negative side gives a deeper, more penetrating arc due to the heat being concentrated in the material being welded. While connecting the electrode to the positive side retains the heat in the electrode which means that the metal in the electrode will burn off quicker depositing the weld material nearer the surface of the weld. This feature makes the connection to the negative side more suited to 'root' welds and connection to the positive side more suitable for 'covering' welds.

Tungsten inert gas (TIG) and metal inert gas (MIG) (aka argon arc welding)

It is found that with some metals, such as aluminium alloys, coated electrodes do not work very well. The coatings cause the aluminium to corrode and, being heavier than the aluminium, slag remains trapped in the weld. It is nevertheless necessary to protect the arc and in such cases an inert gas such as argon may be used for this purpose. Modern equipment uses tungsten rods and argon arc or metal wire welding in each case argon is passed through a tube, down the centre of which is the tungsten electrode. An arc is formed between the work piece and the electrode while the argon forms a shield around the arc. A separate filler wire of suitable material is used to form the joint. The tungsten electrode may be water cooled. This system of welding may be used for most metals and alloys, although care must be taken when welding aluminium as an AC machine may be required. TIG welding is used for welding metals such as aluminium/brass (Yorcalbro), stainless steels and acid-resistant steels.

To master the art of electric welding a person must practice, practice, practice. However, it might be useful to think about some preparation beforehand.

Striking an arc – it may seem obvious but once the welder puts the welding mask in front of his/her face they cannot see the work area until the arc is alight. Therefore, there is very much an element of hand/eye co-ordination coupled with judgement and knowledge which come together to form a mental picture that bridges the gap between being able to see the work with no mask in the way and seeing the weld progress while using the mask.

It is easiest to start with a flat work piece and looking down on the weld. The welder will learn the basic welding process before trying to weld a vertical or overhead weld which are much more difficult.

As the welding progresses the tip of the welding rod must be held in such a way as to keep a constant gap between it and the work piece. However, the rod is melting away; therefore, with the MMA and the TIG welding, the welder must continually adjust the position by moving the holder closer to the work as the length of the welding stick reduces.

With the MIG or wire feed welding the filler material is pushed through the holder and the welder does not have to adjust his/her position relative to the work. In both cases a neat consistent weld is produced only if the speed at progressing the weld is correct.

The angle at which the rod is held relative to the work is also important. If the angle is not correct, then the heat going to the work piece will be uneven and may cut into the parent metal in a way that is not correct for the current weld to maintain strength.

Types of joint and edge preparation

The most efficient method of joining two plates which lie in the same plane is by means of a butt weld, since the two plates then become one continuous section. A square-edge butt (Figure 3.15) may be used for plates up to about 5/6 mm thick. Above this thickness, however, it is difficult to obtain sufficient penetration and it becomes necessary to use a single vee (5–24 mm) (Figure 3.16) or a double vee butt weld (10–30 mm) (Figure 3.17).

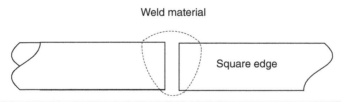

▲ **Figure 3.15** *Square edge*

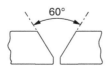

▲ **Figure 3.16** *Single vee*

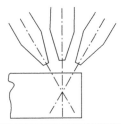

▲ **Figure 3.17** *Double vee*

The double vee welds are more economical as far as the volume of weld metal is concerned, but may require more of the difficult overhead welding and are therefore used only for large thicknesses of plating. The edge preparations for all these joints may be obtained by means of grinding or profile gas or plasma cutting torches having three nozzle burning heads which may be adjusted to suit the required angle of the joint (Figure 3.18).

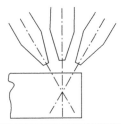

▲ **Figure 3.18** *Cutting heads*

Overlap joints (Figure 3.19) may be used in place of butt welds, but are not as efficient since they do not allow complete penetration of the material and transmit a bending moment to the weld metal. Such joints are used in practice, particularly when connecting brackets to adjacent members. Where these type of joints are allowed the overlap should be more than four time the thickness of the parent metal.

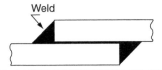

▲ **Figure 3.19** *Overlap*

Fillet welds (Figure 3.20) are used when two members meet at right angles. The strength of these welds depends upon the leg length and the throat thickness, the latter being at least 70% of the leg length. The welds may be continuous on one or both sides of the member or may be intermittent. Continuous welds are used when the joint must be watertight and for other strength members.

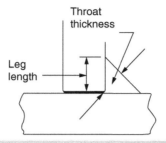

▲ **Figure 3.20** *Fillet weld*

Stiffeners, frames and beams may be connected to the shell plating by intermittent welding (Figure 3.21). In tanks, however, where the rate of corrosion is high, such joints may not be used and it is then necessary to employ continuous welding or to scallop the section (Figure 3.22). The latter method has the advantage of reducing the weight of the structure and improving the drainage, although a combination of corrosion and erosion may reduce the section between the scallops.

▲ **Figure 3.21** *Intermittent welding*

▲ **Figure 3.22** *Scalloped stiffener*

The development of the welded ship

Originally iron and steel ships were put together by overlapping the plating and fixing them with rivets and the very first welded ships followed this technique but used weld instead of rivets. This practice was not very successful; however the science and understanding of 'welded' structures has moved on dramatically since the mid-1940s, so that now all ships are of a welded construction.

The initial change came as the welded construction was much lighter than the equivalent riveted construction, due mainly to the reduction in overlaps and flanges. This meant that a welded ship could carry more cargo for the same loaded draught and warships were faster and could carry more powerful weapons. Welding, if properly carried out, is always watertight without caulking the joints, while in service riveted joints may also be prone to leaks.

With the reduction in overlaps, the outside structure of the ship is also much smoother. This leads to a reduction in hull resistance and coupled with modern hull coatings, gives far better fuel consumption. The smoother surface is easier to clean and less susceptible to corrosion. This has become an important point as the focus on vessels becoming more environmentally friendly has become more acute.

A faulty weld could still, in a modern ship, prove to be dangerous as the structure does not have the 'natural' crack arresting characteristic of the old riveted vessels. However, the up to date quality assurance processes and construction techniques mean that poor welding is not a major issue and where it might happen the quality assurance process will ensure that cracked ships do not enter service without repair. Designers, shipbuilders and surveyors are also knowledgeable about 'stress raisers' and the other causes of cracks, which again means that any defect can be found at an early stage.

The methods of testing welded joints are now successful and much cheaper to carry out than in the past. If a crack starts in a plate it will, under stress, pass through the plate until it reaches the edge. However in the case of the welded ship it could then also carry on and pass through the weld which if left unchecked could prove to be dangerous.

The modern answer to cracking, as in the hulls of older designs, lies in the quality of the design and building of modern ships. The classification societies take the responsibility for the designs of the hull. They specify the standard of the base metal as well as the method and standard of the welds. By taking and testing to destruction test pieces the science behind the welding techniques can be verified. It is then a matter of ensuring that the welders are able to reproduce the required structure consistently to the required design.

Long welds, such as with hull plating, might be carried out using automated welding machines. These may make one, two or three passes to complete the joint. Where this is the case the welds must be checked carefully. This can be completed using ultrasonic non-destructive testing techniques.

It is very important to start with good quality base metals. The International Association of Classification Societies (IACS) gives details of four grades of 'normal' strength steels to be used in shipbuilding and three grades of 'high' strength steels. These are usually manganese steel plates; however in particularly vulnerable areas, such as cargo oil tanks, the International Maritime Organization (IMO) requires that corrosion resistant steels are used. Shipbuilding steel also follows the 'killed steel' process of manufacture. This process removes oxygen from the mix to reduce the porosity of the final product. (See page 66 for more about the chemistry of steel.)

There are still some old vessels that have their hull made of the 'riveted' design. However these are more and more confined to 'historic' ships and some vessels sailing the Great Lakes. In the early welded vessels, surveyors may see a number of longitudinal *crack arrestors* in the main hull structure to reduce the effects of transverse cracks. These crack arrestors may be in the form of riveted seams or strakes of *extra notch tough* steel through which a crack will not pass.

Modern welded designs take great care to reduce the possibility of any cracks forming. This is accomplished by rounding the corners of openings in the structure and by avoiding other stress raising features such as

- the sudden changes in section
- position and design shape of drainage holes
- termination or joining of structural members.

It must be clearly understood, however, that if the cracks appear due to the inherent weakness of the ship, that is, if the bending moment creates unduly high stresses, the crack will pass through the plates whether the ship is riveted, welded or a combination of both.

Cold conditions will have an effect on the steel and classification societies place a minimum performance standard of the specification of the steel used in shipbuilding.

The main structure of a ship can initially be viewed as a large box section or girder. That box section must be fabricated to the dimensions required by the owners to carry the specified cargoes over the required route.

The parts of the structure contributing to the strength calculation of this box section are:

- shell side and bottom plating
- longitudinal stiffeners and bulkheads
- deck plating, longitudinals and large hatch coamings
- inner plating, including any double hull or double bottom constructions
- with passenger ships the accommodation super structure will also contribute to the hull section strength.

The strength of the vessel's hull will also depend upon the thickness and type of steel used in its construction. There will also be a consideration for the proposed lifetime of the vessel. Corrosion will, of course, happen to the vessel while it is in service, and the designer will have to factor this into the strength calculations for the vessel under construction.

High strength steel may have to be used at positions in the design of high stress concentrations. These could be around the construction of hatches or to reduce the high stresses caused by engines, propellers or rough weather (pounding).

Support structures for modern ships have also changed in recent years. The early welded ships required that the ship's senior staff ensure that their vessel was not over-stressed at any time during loading, sailing and discharging.

To assist with this the ship had on board an instrument often referred to as a 'loading computer'. This assisted the staff to understand the stresses and stability for a given set of cargoes. As the vessel was out of contact with shore based support it was the responsibility of the on-board crew to assess the stresses involved.

The structural design of modern ships is such that more sophisticated and accurate methods are required to calculate the loaded stress on a ship before it sets sail. The information required for the senior staff to calculate safe operation comes from the loading manual or the loading computer.

This manual or computer will calculate the still water bending moments and shear forces due to the loads imposed on the ship by the cargo being loaded. This may also relate to the sequencing of loading as ships have been known to fail due to, for example, heavy loads being taken on board fore and aft with nothing in the middle. This has led to bending the hull beyond its construction capabilities leading to the failure.

The loading computers will also be able to give details of the ship's ballast sailing condition and the condition ready for docking. The most modern systems will be able to give 'what if' predictions to help the staff to see what would happen if an area became flooded.

Testing of welds

The testing of weld joints can be divided into two basic types that are fundamentally different. These are:

1. Destructive tests.
2. Non-destructive tests (NDT).

Destructive tests

Specimens of the weld material or welded joint are tested until failure occurs to determine their maximum strength or other characteristics. The tests are standard tests and are the same as those generally used for metals:

1. A tensile test in which the mean tensile strength must be at least 400 MN/m^2.
2. A bend test in which the specimen must be bent through an angle of 90° with an internal radius of 4 times the thickness of the specimen, without cracking at the edges.
3. An impact test in which the specimen must absorb at least 47 J at different temperatures (e.g. −20°C, −10°C, 0°C, +10°C and +20°C). Known as the 'notchy test', it showed up the transition from a ductile to a more brittle state that occurs in steel with a lowering of temperature.
4. Deep penetration electrodes must show the extent of penetration by cutting through a welded section and etching the outline of the weld by means of dilute hydrochloric acid. This test may be carried out on any form of welded joint.

Types of electrode, plates and joints may be tested at regular intervals to ensure that they are maintained at the required standard, while new materials may be checked before being issued or general use. The destructive testing of production work is very useful but limited since it simply determines the strength of the joint before it was destroyed by the removal of the test piece.

Non-destructive tests

An NDT is one of those very useful tools that has seen substantial development over the past few decades.

They usually start with a visual inspection of welded joints which is most important in order to ensure that there are no obvious surface faults, such as cracks and undercut, and to check the leg length and throat thickness of fillet welds. Designers must be careful to ensure that where welds can be completed, there is also sufficient room for them to be visually inspected.

The internal weld structure is now tested with ultrasonic testing techniques. This is an NDT that uses short wavelength sound waves that travel through the metal. The pulsed beams penetrate the metal weld, but if any faults are detected then the waves are disrupted. The changes in the sound are detected and the results analysed by the technicians or displayed on a suitable screen.

Surface cracks which are too fine to be seen even with the aid of a magnifying glass may be outlined using a fluorescent penetrant that enters the crack and may be readily seen with the help of ultra-violet light.

Faults at or near the surface of a weld may be revealed by means of *magnetic crack detection*. An oil containing particles of iron is poured over the weld and then a light electric current is passed through the weld. At the position of any surface faults a magnetic field will be set up which will create an accumulation of the iron particles. Since the remainder of the iron stays in the oil which runs off, it is easy to see where such faults occur.

A more modern system which is being steadily established is the use of *ultrasonics*. A high frequency electric current causes a quartz crystal to vibrate at a high pitch. The vibrations are transmitted directly through the material being tested. If the material is homogeneous, the vibration is reflected from the opposite surface, converted to an electrical impulse and indicated on an oscilloscope. Any fault in the material, no matter how small, will cause an intermediate reflection which may be noted on the screen. This method is useful in that it will indicate a lamination in a plate which will not be shown on the older X-ray method of testing. Ultrasonics are now also being used to determine the thickness of plating in repair work and to avoid the necessity of drilling through the plate.

The rules of IACS state that any person carrying out a visual inspection must have sufficient knowledge and experience and any person undertaking magnetic particle testing or liquid penetrant testing must be qualified to the IACS standards. Such persons will also have certification to prove their qualifications.

IACS gives further information about the rules for testing and repair of crankshafts, propeller shafts and rudder stocks.

Magnetic particle inspection works on the principle that a defect in the metal, which could be below the surface of the metal, causes a distortion in a magnetic flux at the point of the defect. The distortion of the magnetic flux is disproportionately large and extends to the surface of the metal. This leakage of magnetic flux can then be used to attract coloured iron particles suspended in a solution.

Liquid penetrant testing is a low-cost method of testing for cracks that break the surface of a non-porous material. It relies on the penetrating ability of a low surface tension fluid such as paraffin to carry a dye into a surface crack that may not be visible under normal circumstances.

Ultrasonic testing is now commonly associated with discussions about NDT. However, this is more a technique for detecting the thickness of metal and internal defects, than it is to determine the extent of any corrosion taking place. This method of testing has the added advantage of being open to a high degree of automation and, therefore, slightly less training is needed to carry out the tests successfully. It also lends itself well to testing for welding defects

Faults in welded joints

Electric welding, using correct technique, suitable materials and conditions, should produce faultless welds. Should these requirements not be met, however, faults will occur in the joint (Figure 3.23). If the current is too high, the edge of the plate may be burned away. This is known as *undercut* and has the effect of reducing the thickness of the plate at that point. It is important to chip off all of the slag, particularly in multi-run welds, otherwise *slag inclusions* occur in the joint, again reducing the effective thickness of the weld. The type of rod and the edge preparation must be suitable to ensure *complete penetration* of the joint. In many cases a good surface appearance hides the lack of fusion beneath, and, since this fault may be continuous in the weld, could prove very dangerous. Incorrect welding technique sometimes causes bubbles of air to be trapped in the weld. These bubbles tend to force their way to the surface leaving *pipes* in the weld. Smaller bubbles in greater quantities are known as *porosity*. *Cracks* on or below the surface may occur due to unequal cooling rates or an accumulation of weld metal. The rate of cooling is also the cause of distortion in the plates, much of which may be reduced by correct welding procedure.

Another fault which is attributed to welding but which may occur in any thick plate, especially at extremely low temperatures, is *brittle fracture*. Several serious failures occurred during and just after World War II, when large quantities of welded work were produced. Cracks may start at relatively small faults and suddenly pass through the plating at comparatively small stresses. It is important to ensure that no faults or discontinuities occur, particularly with respect to important structural members. The grade of steel used must be suitable for welding, with careful control of the manganese/carbon content in the greater thicknesses to ensure notch-tough qualities.

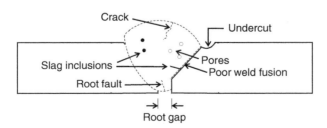

▲ **Figure 3.23** *Weld defects*

Design of welded structures

It was very quickly realised that welding is different from riveting not only as a process, but as a method of attachment. It is not correct to just update a riveted structure by welding it, and indeed the whole shipyard had to change to accommodate the new methods.

A greater continuity of material is obtained with a welded structure resulting in more efficient designs. Many of the faults which occurred in welded ships were due to the large number of components that were welded together resulting in high stress points which caused fractures.

The following illustration serves well to show students the major differences with modern welded structures. Consider the structure of an old oil tanker. Figure 3.24 shows part of a typical 'riveted' centre girder, connected to a vertical bulkhead web. Initially when such ships were built of welded construction, the same type of design was used with the riveting being replaced by welding, resulting in the type of structure shown in Figure 3.25. Note that the 'overlap' to accommodate the rivets has been removed under the new process.

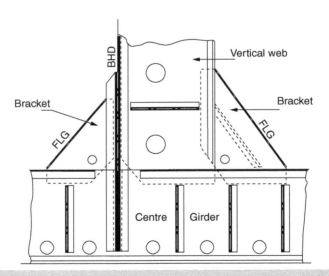

▲ **Figure 3.24** *Structure of old oil tanker*

However it was found with such designs that cracks occurred at the toes of the brackets and at the ends of the flats. The brackets were then built in to the webs using continuous face flats having small radii at the toes of the brackets. Cracks still appeared showing that

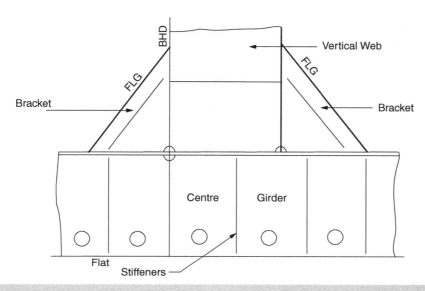

▲ **Figure 3.25** *A more recent example*

the curvature was too small. The radii were increased until eventually the whole bracket was formed by a large radius joining the bottom girder to the vertical web (Figure 3.26). This type of structure is now regarded as commonplace in modern ship tanker designs.

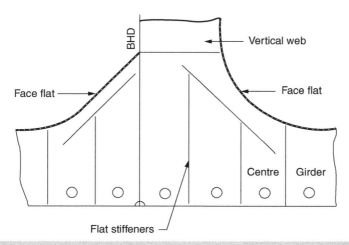

▲ **Figure 3.26** *Modified to remove stress raisers*

Great care must be taken to ensure that structural members on opposite sides of bulkheads are perfectly in line, otherwise cracks may occur in the plating due to shearing.

Welded structures are shown in the succeeding chapters dealing with modern ship construction.

Materials

Mild steel

Mild steel or low carbon steel in several grades has been used as a structural material for shipbuilding for over a century. It has the advantage of having a good, cost-effective, strength–weight ratio. It is used in the construction of all types of ships especially for the construction of the hull. Some types such as superyachts do use alternative materials such as glass reinforced plastic (GRP).

Steel is the basic combination of iron and carbon. Mild steel contains up to about 0.25% carbon. Medium carbon steels have about 0.25–0.45% carbon and high carbon steel has between 0.45% and 1.50%. Cast iron has a carbon content of between 2.5% and 4.5%. With shipbuilding steels the oxygen is removed before solidification by adding manganese to the mix. These steels are then termed 'killed' steels. With the permission of a classification society, grade 'A' steel used can be of the 'rimmed' variety which is where a small amount of the oxygen is left in the chemical mix.

There are four grades of steel in common use in shipbuilding. The manufacturing process and chemical composition are carefully specified by the classification societies and the steels are known as IACS steels. Class also requires the results of tests to be included in any 'approval' process for the manufacturer. These tests will include tensile tests, Charpy V-notch (CVN) impact tests, weldability tests as well as tests carried out on welded specimens.

Approved manufacturers will also have to provide evidence of how they certify their steel plate for chemical composition, deoxidation, fine grain and thickness tolerance as well as any heat treatment carried out on the metal.

The steels are grades A, B, D and E, and they vary in their chemical composition and therefore in their strength and degree of CVN toughness. Grade A has the least resistance to brittle fracture while Grade E is termed 'extra notch tough'. Grade D has sufficient resistance to cracks for it to be used extensively for main structural material.

The disposition of the grades in any ship depends upon the thickness of the material, the part of the ship under consideration and the stress to which it may be subject. For example, the bottom shell plating of a ship within the midship portion of the ship will have the following grade requirements.

Plate thickness	Grade of steel
Up to 20.5 mm	A
20.5–25.5 mm	B
25.5–40 mm	D
Above 40 mm	E

The normal strength steel (A–E) has a 'yield strength' of 235 MN/m^2 and the tensile strength of the different grades remains constant at between 400 MN/m^2 and 520 MN/m^2. The difference lies in the chemical composition which improves the impact strength of D and E steels. Impact resistance is measured by means of a Charpy test in which specimens may be tested at a variety of temperatures. The following table shows the minimum values required by Lloyd's Register.

Type of steel	Temperature	Impact resistance
B	0°C	27 joules
D	0°C	47 joules
E	−40°C	27 joules

Higher tensile steels

As oil tankers, bulk carriers and now container ships increased in size the thickness of steel required for the main longitudinal strength members also increased. In an attempt to reduce the thickness of material and thereby reduce the light displacement of the ship, classification societies accept the use of steels of higher tensile strength. These steels are designated AH, BH, DH and EH and may be used to replace the normal grades for any given structural member. Thus a bottom shell plate amidships may be 30 mm in thickness of grade DH steel.

The yield strength of the higher grade steel is between 315 MN/m^2 and 390 MN/m^2 with the tensile strength being increased to between 490 MN/m^2 and 620 MN/m^2 and having the same percentage elongation as the low carbon steel. Thus it is possible to form a structure combining low carbon steel with the more expensive, but thinner higher tensile steel. The latter is used where it is most effective, that is, for upper deck plating and longitudinals, and bottom shell plating and longitudinals.

Care must be taken in the design to ensure that the hull has an acceptable standard of stiffness, otherwise the deflection of the ship may become excessive. Welding must be carried out using low hydrogen electrodes, together with a degree of preheating. Subsequent repairs must be carried out using the same type of steel and electrodes. It is a considerable advantage if the ship carries spare electrodes, while a plan of the ship should be available showing the extent of the material together with its specification.

Arctic D steel

If part of the structure of a ship is liable to be subject to particularly low temperatures, then the normal grades of steel are not suitable, as they are susceptible to 'brittle fracture'. A special type of steel, known as Arctic D, has been developed for this purpose. It has a higher tensile strength than normal mild steel, but its most important quality is its ability to absorb a minimum of 40 J at −55°C in a Charpy impact test using a standard specimen. The International Association of Classification Societies (IACS) also require vessels that conform to the 'Polar Code' to have the structural parts (that are in contact with the water) to be constructed of steel that is manufactured to the correct class. The Charpy test for these steels will need to be completed at −10°C and achieve an absorption of 20 J of energy.

Aluminium alloys

Pure aluminium is too soft for use as a structural material and must be alloyed to provide sufficient strength in relation to the mass of material used. The aluminium is combined with copper, magnesium, silicon, iron, manganese, zinc, chromium and titanium, the manganese content varying between about 1% and 5% depending upon the alloy. The alloy must have a tensile strength of 260 MN/m^2 compared with 400–490 MN/m^2 for mild steel.

There are two major types of alloy used in shipbuilding, heat-treatable and non-heat-treatable. The former is heat-treated during manufacture and, if it is subsequently heat-treated, tends to lose its strength. Non-heat-treatable alloys may be readily welded and subject to controlled heat treatment while being worked.

The advantages of aluminium alloy in ship construction lie in the reduction in weight of the material and its non-magnetic properties. The former is only important, however, if sufficient material is used to significantly reduce the light displacement of the ship and hence increase the available deadweight or reduce the power required for any given deadweight and speed. Unfortunately, the melting point of the alloy (about 600°C) lies well below the requirements of a standard fire test maximum temperature (927°C). Thus if it is to be used for fire subdivisions it must be suitably insulated.

The major application of aluminium alloys as a shipbuilding material is in the construction of passenger ships, where the superstructure may be built wholly of the alloy. The saving in weight at the top of the ship reduces the necessity to carry permanent ballast to maintain adequate stability. The double saving results in an economical justification for the use of the material. Great care must be taken when attaching the aluminium superstructure to the steel deck of the main hull structure (see Chapter 12).

Other applications in passenger ships have been for cabin furniture, lifeboats and funnels.

One tremendous advantage of aluminium alloy is its ability to accept impact loads at extremely low temperatures. Thus it is an eminently suitable material for main tank structure in low-temperature gas carriers.

The fracturing of steel structures

When welding was first introduced into shipbuilding on an extensive scale, several structural failures occurred. Cracks were found in ships which were not highly stressed, indeed in some cases the estimated stress was particularly low. On investigation it was found that the cracks were of a brittle nature, indicated by the crystalline appearance of the failed material. Further studies indicated that similar types of fault had occurred in riveted ships although their consequences were not nearly as serious as with welded vessels.

Series of tests indicated that the failures were caused by brittle fracture of the material. In some cases it was apparent that the crack was initiated from a notch in the plate: a square corner on an opening or a fault in the welding. (In 1888 Lloyd's Register pointed out the dangers of square corners on openings.) At other times cracks appeared suddenly at low temperatures while the stresses were particularly low and no structural notches appeared in the area. Some cracks occurred in the vicinity of a weld and were attributed to the change in the composition of the steel due to welding. Excessive impact loading also created cracks with a crystalline appearance. Explosions near the material caused dishing of thin plate but cracking of thick plate.

The consequences of brittle fracture may be reduced by fitting crack arrestors to the ship where high stresses are likely to occur. Riveted seams or strakes of extra notch tough steel used to be fitted in the decks and shell of large tankers and bulk carriers.

In modern vessels the problem of brittle fracture may be reduced or avoided by designing the structure so that notches in plating do not occur, and by using steel which has a reasonable degree of notch-toughness. Grades D and E steel lie in this category and have proved very successful in service for the main structure of ships where the plates are more than about 12 mm thick.

Much more is now known about the stresses that are set up in different parts of a ship's structure. Great care is taken over getting constructional details correct. Care must be taken with the quality of the welding. Sharp corners act to raise the stress levels within the surrounding steel plate. Changing from one framing system to the other is not straightforward.

Composites

Composite materials are materials that have been built up in different ways to form the structure required. Different combinations have different properties. The quality of the finished product is very much dependent upon the care, expertise and experience of the manufacturer. Normally the term refers to glass reinforced fibre polymers (GRP) or to carbon reinforced fibre polymers (CRP) or Kevlar.

The main advantage of composite materials is their strength-to-weight ratio which can be many times higher than steel. In the marine environment especially another potential advantage is that they are non-corrosive and therefore not affected by sea water. One of the areas where composites have impacted upon the most is with the hull of superyachts.

Composite materials are not however a wonder cure to be used in every eventuality. Engineering materials have strengths and weaknesses and composite materials are no exception. They may be brittle under some conditions and may not show the same characteristics when force is applied in different directions.

Composites can also become porous, and osmosis is sometimes a problem with GRP hulls. Manufacturing defects can also cause structural weaknesses. Within the marine field composites are gaining popularity for the construction of pipework within machinery spaces.

Machining and Repairs

Part of the manufacturing of a ship involves machining some of the materials, in order to fit the application they were designed for. One of the most obvious items that will require machining is the propeller.

Other items could be the foundations for engine pumps and power transmission shaft bearings. However, together with welding, machining metal components will mostly take place during the process of carrying out necessary repairs to the equipment while the ship is away from suitable repair facilities ashore.

Machine tools have the ability to perform to high levels of accuracy. Higher levels of vibration onboard ships, however, mean that it is not always possible to achieve the best results. Calmer conditions result from the ship being in port, due to the main engines not being in operation. The process of using the machines will also be safer in port due to less movement of the vessel from the effects of being in the water.

The most common machine tools found on-board are;

- centre lathes
- pillar drills
- offhand grinding wheels (abrasive wheels)
- milling machines
- shaping machine.

Centre lathes

The main observation with centre lathes is that they rotate the workpiece and are arranged to have the cutting tool fixed. This means that a good level of knowledge and skill is required by the operator if accidents are to be avoided. The main features of the centre lathe are shown in Figure 3.27.

It should be noted that all operators of centre lathes should only work with the safety guards fitted correctly and in good working order. It is also important to switch off the power to the machine when the machine is left unattended. When in general use, the centre lathe will be fitted with a three-jaw, self-centering chuck that is used to hold the workpiece.

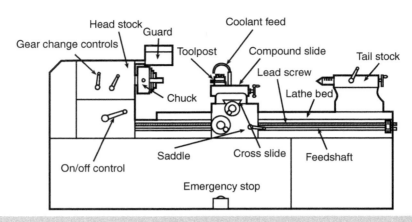

▲ **Figure 3.27** *Centre lathe*

This chuck is less accurate that the 'four jaw' chuck but it has the advantage of all the jaws being operated together, when any one of the adjusting screws are turned by the chuck key.

The three jaws move in/out together toward or away from the centre of rotation of the chuck. This means that metal shapes that are symmetrical about a centre line can be griped and machined relatively easily as the chuck automatically centres the workpiece.

More complicated shaped work or where a piece must be machined, off centre, then the use of a four-jaw chuck is required. Here the chuck has four jaws that are adjusted 'independently' and can therefore be at different distances from the centre line of the chuck/lathe. This means that the centre of rotation of the work need not be its centre line.

There are several real dangers to note with using the centre lathe, these are:

- leaving the chuck key inserted in the adjustment hole when the machine is switch on, causing the key to fly out due to the centrifugal force exerted as the chuck starts to move;
- incorrect positioning of the cutting tool in relation to the work piece;
- cutting tool angles set incorrectly;
- incorrect rotational speed for the type of metal being cut;
- work piece extending too far from the jaws and, therefore, not being held ridgedly enough;
- all safety guards covering the rotating components and the cutting tool need to be used at all times;
- metal shavings, known as 'swarf', should not be removed with hands, as these are very sharp;
- loose clothing, especially baggy sleeves, must be tied out of the way.

Pillar drill

The pillar drill will probably be the most used machine tool on board. Drill operators will need to be very careful when operating this machine, but they will not need the same level of training as is required by the operators of the centre lathe. With this, straightforward machine, the workpiece is held rigid with strong clamping arrangements while the cutting tool (drill) is doing the rotating. The main features of the pillar drill are shown in Figure 3.28.

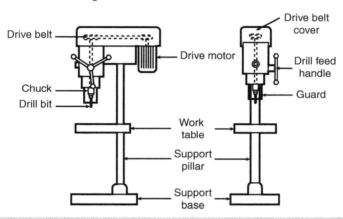

▲ **Figure 3.28** *Pillar drill*

The important actions when using this machine tool are;

- adequate clamping of the workpiece
- using the guard around the cutting edge of the drill bit
- using a sharp drill
- correct speed for the size of drill and type of metal of the work piece
- keep loose clothing from coming into contact with the drill bit.

Offhand grinding wheel (abrasive wheel)

This machine tool is usually arranged with two wheels, one on either end of a spindle that is driven from the centre by an electric motor. The abrasive wheels fitted are used to reduce metal shapes by grinding off sections. The main features of the offhand grinding wheel are shown in Figure 3.29.

The metal to be ground, is held against the rest and pushed against the wheel that then cuts into the work piece. Grinding wheels are dangerous and need to be used by people who have received guidance from a 'competent' person.

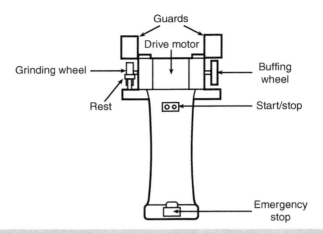

▲ **Figure 3.29** *Offhand grinding wheel*

The International Labour Organization (ILO) sets out the safety practices that should be followed in their booklet 'Safety and Health in Shipbuilding and Ship Repair'.

The abrasive wheels are manufactured from different materials that are pressed together. Within the composition there are very hard and sharp abrasive materials, moulded together with a binding material such as a resin or, more commonly, glass, in the so called 'vitrified' bond.

The wheel works by the 'sharp' (usually silicon carbide or aluminium oxide) particles acting as miniature cutting tools reducing the workpiece bit by bit. Eventually, the top layers are removed, uncovering further 'sharp' particles. The sharp particles fly off as hot sparks and care should be taken to avoid any fire risk.

The wheels are porous and, therefore, can be very dangerous if water is allowed to soak into the wheel. The centrifugal force on the trapped water when the wheel starts to rotate can be enough to shatter the structure, sending out very sharp objects at high speed.

Milling and shaping machines

These are useful, but more specialised machines, and are not found on board many ships. The milling machine would be used in a vibration-free environment, for cutting groves such as the keyway in a transmission shaft, or the drive slot in a faceplate.

Due to the environment on board, the milling machine is restricted in its ability to undertake an effective role and, therefore, will not often be seen on ships.

It is a similar situation with the 'shaping machine'. When ships were away from technical support for long periods of time and onboard staff were required to undertake a higher degree of manufacturing processes than is required on a modern vessel, the shaping machine had more uses than is currently the case.

4

BOTTOM AND SIDE FRAMING

Double Bottom

Ocean-going ships (with the exception of tankers, which now have to be double hulled) and most coastal vessels are fitted with a double bottom system of construction, which extends from the fore peak bulkhead almost to the after peak bulkhead.

The double bottom consists of the outer shell and an inner skin or tank top between 1 m and 1.5 m above the keel. This provides a form of protection in the event of damage to the bottom shell, and it also provides protection to the environment from any oil or contaminants that may be in the bilges at the time of a breach of the hull. However, the International Convention for the Prevention of Pollution from Ships (MARPOL) specifies whether, and how, fuel and lubrication oil is permitted to be stored in 'double bottom' tanks.

The tank top, being continuous, contributes to the hull girder strength. From 1997 additional support was required for the platforms around the cargo and machinery areas. These added strength items are called 'solid floors' and are thicker than the normal double bottom plating.

Additional strength may also be required for high powered engines or gearboxes and thrust blocks bolted directly to the bottom plating will need to be fixed to plating of at least 19 mm thickness.

Designers must also indicate on their plans exactly how the docking loads are to be accommodated. Additional brackets or high strength material might have to be used to withstand the high 'local' loading that will need to be transmitted to support the vessel during this time.

The double bottom space contains a considerable amount of scantlings and is therefore unsuitable for carrying much cargo. Where the regulations allow, double bottom tanks may be used for the carriage of oil fuel, fresh water and water ballast. They are subdivided longitudinally and transversely to reduce any free surface effect.

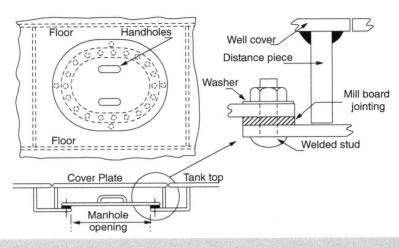

▲ **Figure 4.1** *Well-type manhole*

Double bottom tanks can be filled or emptied with the different liquids that are required to be carried, and they can also be used to correct the heel of a ship or to change the trim. Access to these tanks is arranged in the form of manholes with watertight covers (Figure 4.1) and care must be taken when entering these tanks as they are dangerous spaces and could have an oxygen-depleted or poisonous atmosphere.

In the majority of ships only one watertight longitudinal division, a centre girder, is fitted, but many modern ships are designed with either three or four tanks across the ship.

During the 1970s and 1980s designers started to produce roll-on roll-off car ferries and cargo ships. The problem was that the main ro-ro deck was continuous from bow to stern and had large doors at either end. In general, seafarers realised that only a very shallow depth of water across that main deck would be sufficient to capsize the vessel. Following a major disaster, where that actually happened, the rules were changed and watertight divisions were included in the design of these vessels to reduce the 'free surface effect' in the event of any water making its way onto the main deck.

A cofferdam must be fitted between a fuel tank and a fresh water tank to prevent contamination of one with the other, if a failure occurs in one tank. The tanks are tested by pressing them up until they overflow. Since the overflow pipe usually extends above the weather deck, the tank top is subject to a tremendous head which in most cases will be sufficient as a test for water tightness and will be greater than the weight of the cargo pressing down from the hold.

The tank top plating must be thick enough to prevent undue distortion when the cargo is loaded. In bulk carriers, if it is anticipated that cargo will be regularly discharged by grabs or by forklift trucks, it will be necessary to fit either additional protection to the ceiling of the tank or heavier flush plating. Under hatchways, where the tank top is most liable to damage, the plating or protection must be increased to the tank's ceiling, and the plating is at least 10% thicker in the engine room.

In the lower part generally considered to be the bilges, the tank top may be either continued straight out to the shell, or knuckled down to the shell by means of a tank margin plate set at an angle of about 45° to the tank top and meeting the shell almost at right angles.

It has the added advantage, however, of forming a bilge space into which water may drain and has, in the past, proven to be popular. If no margin plate is fitted it is necessary to fit drain hats or wells in the after end of the tank top in each compartment so that the bilges can be pumped dry.

Internal structure of the hull

The hull girder strength can be enhanced with the inclusion of a continuous *centre girder* and/or *side girders*, these extending longitudinally from the fore peak to after peak bulkhead. The centre girder is usually watertight except at the extreme fore and after ends where the ship is narrow, although there are some designs of ship where the centre girder does not form a tank boundary and is therefore not watertight. From 1997 onwards the plates making up the *0.75L* midship section are to be continuous. Centre girders must also provide sufficient strength to withstand the docking loads and additional 'docking brackets' may need to be included in the design.

A pipe tunnel may be substituted for a centre girder as long as the construction of the tunnel is of sufficient strength. If the ship's designer wishes to include this feature, a full set of detailed plans must be submitted for approval.

Additional longitudinal *side girders* are fitted (at a maximum of 5 m apart) depending upon the breadth of the ship but these are neither continuous nor watertight, having large manholes or lightening holes in them. Special consideration must be given to providing side girders under the machinery space and/or the thrust block seating.

The tanks are divided transversely by *watertight floors*, which in most ocean-going ships are required to be stiffened vertically, to withstand the liquid pressure. Figure 4.2 shows a typical watertight floor.

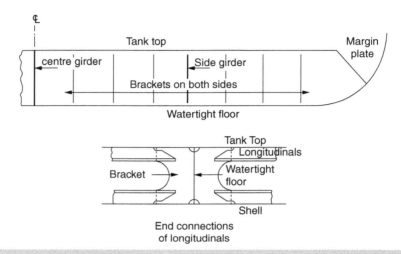

▲ **Figure 4.2** *Watertight floor and end connections of longitudinal*

In ships less than 120 m in length the bottom shell and tank top are supported at intervals of not more than 3 m by transverse plates known as *solid floors*. The name slightly belies the structure since large lightening holes are cut in them. In addition, small air release and drain holes are also cut at the top and bottom, respectively. These holes are most important since it is essential to have adequate access and ventilation to all parts of the double bottom. There have been many cases of personnel entering tanks which have been inadequately ventilated, with resultant gassing or suffocation. *These tanks must still be regarded as enclosed spaces.*

The solid floor is usually fitted as a continuous plate extending from the centre girder to the margin plate. The side girder is therefore broken on each side of the floor plate and is referred to as being *intercostal*.

Solid floors are required at every frame space in the machinery room, in the forward quarter length and elsewhere where heavy loads are experienced, such as under bulkheads and boiler bearings.

The remaining bottom support may be either of two forms:

(a) transverse framing

(b) longitudinal framing.

Transverse framing was the more traditional method of ship construction which followed on from the methods used on riveted ships, and is still a useful method used in the construction of smaller vessels. The shell and tank top between the widely spaced solid floors are stiffened by bulb angles or similar sections running across the ship and attached at the centreline and the margin to large flanged brackets. Additional support is given to these stiffeners by the side girder and by intermediate struts which are fitted to reduce the span. Such a structure was known as a bracket floor and is still referred to by some classification society rules for the construction of a ship's hulls (Figure 4.3).

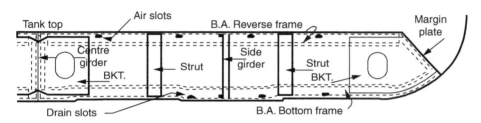

▲ **Figure 4.3** *Bracket floor*

Buckling caused by the distortion due to the welding of the floors and frames, together with the bending of the ship, needs to be guarded against and designers are now required to specify longitudinal stiffening in the double bottom for all ships over 120 m long.

Longitudinal frames are fitted to the bottom shell and under the tank top, at intervals of about 760 mm. They are supported by the solid floors mentioned earlier, although the spacing of these floors may be increased to 3.7 m. Intermediate struts are fitted so that the unsupported span of the longitudinals does not exceed 2.5 m. Brackets are again required at the margin plate and centre girder, the latter being necessary when docking as mentioned earlier. Figure 4.4 shows an arrangement of a double bottom construction.

The longitudinals are then arranged to line up with any additional longitudinal girders which are required for machinery or thrust block support.

Duct keel or pipe tunnels

Some ships might still be fitted with a tunnel or tunnels which is a convenient method of routing any pipework that is required to supply services to the forward part of the vessel. These are known as duct keels (see Figure 4.5) and as long as they are of equal strength, they can be used in place of the centre girder explained on pages 42 and 77. The pipe tunnel extends from within the engine room along the length of the vessel to the forward holds. This arrangement then allows the pipes to be carried beneath the hold spaces and are thus protected against cargo damage. Access into the duct is arranged from the engine room. The pipes can then be inspected and repaired at any time independent of the weather (within working constraints) and cargo operations.

At the same time it is possible to carry oil and water pipes in the duct, preventing contamination which could occur if the pipes passed through tanks. Duct keels are particularly important in insulated ships, allowing access to the pipes without disturbing the insulation. Ducts are not required aft since the pipes may be carried through the shaft tunnel.

The duct keel is formed by two longitudinal girders up to 1.83 m apart. This distance must not be exceeded as the girders must be supported by the keel blocks when docking. The structure on each side of the girders is the normal double bottom arrangement. The keel and the tank top centre strake must be strengthened either by supporting members in the duct or by increasing the thickness of the plates considerably.

It is vital that the duct space is treated as an enclosed space and great care must be taken before any person enters the area.

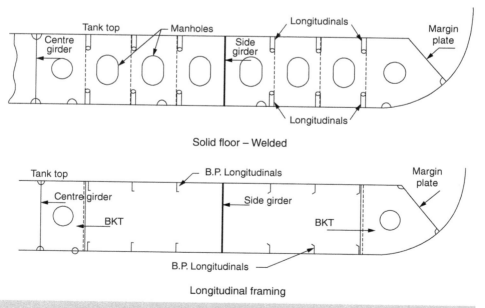

Solid floor – Welded

Longitudinal framing

▲ **Figure 4.4** *Solid floor – welded and longitudinal framing*

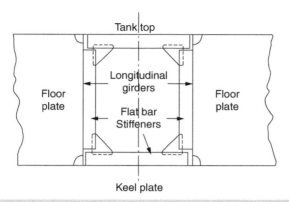

▲ **Figure 4.5** *Duct keel*

Double bottom in the machinery space

Great care must be taken in the machinery space to ensure that the main and auxiliary machinery are efficiently supported. Weak supports may cause damage to the machinery, while large unsupported panels of plating may lead to vibration of the structure.

Where the main engine is a large two stroke or medium speed four stroke diesel the bedplates are bolted through a tank top plate which is to be at least 19 mm thick and is continuous to the thrust block seating. Designers may need to provide detailed engineering analysis of the structure in this area especially if very high power engines are used.

Diesel electric ships may have a larger number of smaller engines that could then be situated higher than near the double bottom construction. In this case the designers will need to consider the strength of the floors under these engines and additional girders might be used to meet this requirement.

A girder is fitted on each side of the bedplate in such a way that the holding down bolts pass through the top angle of the girder. In welded ships a horizontal flat is sometimes fitted to the top of the girder with respect to the holding down bolts (Figure 4.6).

In motor ships where a drain tank is required under the machinery, a cofferdam is fitted giving access to the holding down bolts and isolating the drain from the remainder of the double bottom tanks. Additional longitudinal girders are fitted with respect to heavy auxiliary machinery such as generators.

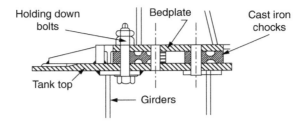

▲ **Figure 4.6** *Main engine holding down bolts*

Note the cast iron chocks in Figure 4.6. More up-to-date arrangements use 'resin' chocks, and resilient mountings are used on small or medium sized engines. The resilient mountings are able to soak up more vibration from the engines, thus making the machinery space and the ship quieter. Resilient mountings can be manufactured from rubber and may also incorporate some form of springing.

Side Framing

The side shell is supported by frames which run vertically from the tank margin to the upper deck. These frames, which are spaced about 760 mm apart, are in the form of bulb angles and channels in riveted ships, and are bulb plates/flats (Figure 3.4) in welded ships. The lengths of frames are usually broken at the decks, allowing smaller sections to be used in the 'tween deck spaces where the load and span are reduced. The hold frames are of large section (300 mm bulb angle). They are connected at the tank margin to flanged tank side brackets (Figure 4.7). To prevent the free edge of the brackets buckling, a gusset plate is fitted, connecting the flange of the brackets

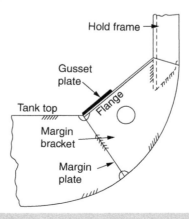

▲ **Figure 4.7** *Tank side bracket*

to the tank top. A hole is cut in each bracket to allow the passage of bilge pipes. In insulated ships the tank top may be extended to form the gusset plate and the tank side bracket fitted below the level of the tank top (Figure 4.8). This increases the cargo capacity and facilitates the fitting of the insulation. Since the portion of the bracket above the tank top level is dispensed with, the effective span of the frame is increased, causing an increase in the size of the frame.

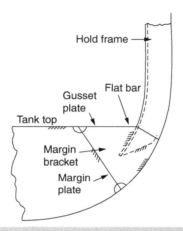

▲ **Figure 4.8** *Tank side bracket insulated ship*

The tops of the hold frames terminate below the lowest deck and are connected to the deck by beam knees (Figure 4.9) which may be flanged on their free edge to give added stiffness and strength. The bottoms of the 'tween deck frames are usually welded directly to the deck, the deck plating at the side being knuckled up to improve drainage. At the top, the 'tween deck frames are stopped slightly short of the upper deck and connected by beam knees (Figure 4.10). In some cases, the 'tween deck

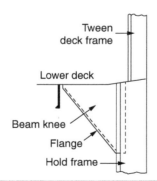

▲ **Figure 4.9** *Lower deck beam knee*

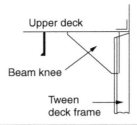

▲ **Figure 4.10** *Upper deck beam knee*

frames must be carried through the second deck and it is necessary to fit a collar round each frame to ensure that the deck is watertight. Figure 4.11 shows a typical collar arrangement, the collar being in two pieces, welded right round the edges.

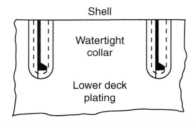

▲ **Figure 4.11** *Welded plate collars*

Wood sparring or similar protective lining can be fitted to the toes of the hold and 'tween deck frames to protect the cargo from damage, while the top of the tank side brackets in the holds are fitted with wood ceiling.

Web frames may be fitted in the machinery space and connected to strong beams or pillars in an attempt to reduce vibration (Figure 4.12). These web frames are about 600 mm deep and are stiffened on their free edge. It is usual to fit two or three web frames on each side of the ship, a smaller web being fitted in the 'tween decks. The exact scantling requirements will be determined by the strength calculations completed by the designer.

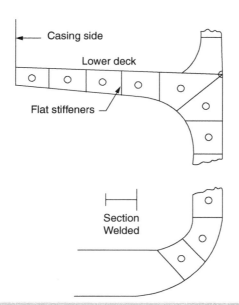

Casing side

Lower deck

Flat stiffeners

Section
Welded

▲ **Figure 4.12** *Web frame*

5

SHELL AND DECKS

The external hull of a ship consists of bottom plates making up the bottom of the shell. Plates making up the side shell and the decks are formed into longitudinal strips of plating known as strakes. The strakes themselves are constructed of a number of plates joined end to end and large, wide plates are used to reduce the welding required. Invariably the hull of a modern ship is built up in 'pre-fabricated' sections. Modern measuring and fabrication accuracy has enabled these 'pre-fabricated' sections to be assembled remotely and brought together in the final construction. Strict adherence to the class rules relating to manufacturing techniques will produce the necessary accuracy required. These sections can even be built up remotely from the main site and the sections brought together to make up the final hull.

Shell Plating

The bottom and side shell plating of a ship also forms a major part of the longitudinal strength of the vessel. The most important part of the shell plating is that on the bottom of the ship, since this is the greatest distance from the neutral axis. As it is subjected to the highest forces, it is slightly thicker than the side shell plating.

The *keel plate* is about 30% thicker than the remainder of the bottom shell plating, since it is subject to additional wear and tear when dry-docking. The strake adjacent to the keel on each side of the ship is known as the *garboard strake* which is the same thickness as the remainder of the bottom shell plating. The uppermost line of plating in the side shell is known as the *sheerstrake* which is 10–20% thicker than the remaining side shell plating.

The thickness of the shell plating depends mainly on the length of the ship, varying between about 10 mm at 60 m and 20 mm at 150 m. The depth of the ship, the maximum draught and the frame spacing are, however, also taken into account. If the depth is increased, it is possible to reduce the thickness of the plating. In ships fitted with long bridges which extend to the sides of the ship, the depth with respect to the bridge is increased, resulting in thinner shell plating. Great care must be taken at the ends of such superstructures to ensure that the bridge side plating is tapered gradually to the level of the upper deck, while the thicker shell plating forward and aft of the bridge must be taken past the ends of the bridge to form an efficient scarph. If the draught of the ship is increased, then the shell plating must also be increased. Thus a ship whose freeboard is measured from the upper deck has thicker shell plating than a similar ship whose freeboard is measured from the second deck. If the frame spacing is increased the shell plating is required to be increased.

The maximum bending moment of a ship occurs at or near amidships. Therefore the shell plating must be of sufficient strength to ensure its contribution to the hull girder strength, and it is reasonable to build the ship stronger amidships than at the ends. The main shell plating has its thickness maintained for 40% of its length amidships and tapered *gradually* to a minimum thickness at the ends of the ship.

While the longitudinal strength of shell plating is of prime importance, it is equally important to ensure that its other functions are not overlooked. Watertight hulls were made before longitudinal strength was considered. It is essential that the shell plating should be watertight and, at the same time, capable of withstanding the static and dynamic loads created by the water. The shell plating, together with the frames and double bottom floors, resist the water pressure, while the plating must be thick enough to prevent undue distortion between the frames and floors. If it is anticipated that the vessel will regularly travel through ice, the shell plating in the region of the waterline forward is increased in thickness and small intermediate frames are fitted to reduce the widths of the panels of plating. The bottom shell plating forward is increased in thickness to reduce the effects of pounding (see Chapter 7).

The shell plating and side frames act as pillars supporting the loads from the decks above and must be able to withstand the weight of the cargo. In most cases the strength of the panel which is required to withstand the water pressure is more than sufficient to support the cargo, but where the internal loading is particularly high, such as with respect to a deep tank, the frames must be increased in strength.

It is necessary on exposed decks to fit some arrangement to prevent personnel falling or being washed overboard. Many ships are fitted with open rails for this purpose while others are fitted with solid plates known as *bulwarks* at least 1 m high. These bulwarks are much thinner than the normal shell plating and are not regarded as longitudinal strength members. The upper edge is stiffened by a 'hooked angle', that is, the plate is fitted inside the flange. This covers the free edge of the plate and results in a neater arrangement. Substantial stays must be fitted from the bulwark to the deck at intervals of 1.83 m or less. The lower edge of the bulwark in riveted ships was fixed to the top edge of the sheerstrake. In welded ships, however, there must be no direct connection between the bulwark and the sheerstrake, especially amidships, since the high stresses would then be transmitted to the bulwark causing cracks to appear. These cracks could then pass through the sheerstrake.

Large openings, known as *freeing ports*, must be cut in the bottom of the bulwark to allow the water to flow off deck when a heavy sea is shipped. Failure to clear the water could cause the ship to capsize. Rails or grids are fitted to restrict the opening to 230 mm in depth, while many ships are fitted with hinged doors on the outboard side of the freeing port (see Figure 5.1), acting as rather inefficient non-return valves. It is essential that there should be no means of bolting the door in the closed position.

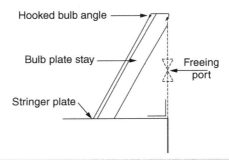

▲ **Figure 5.1** *Bulwark*

Deck Plating

The deck plating of a ship carries a large proportion of the stresses due to longitudinal bending, the upper deck carrying greater loads than the second deck. The continuous plating alongside the hatches must be thick enough to withstand the loads. The plating between the hatches has little effect on the longitudinal strength. The thickness of plating depends largely upon the length of the ship and the width of deck alongside the hatchways. In narrow ships, or in vessels having wide hatches, the thickness of plating is increased. At the ends of the ship, where the bending moments are reduced, the thickness of plating may be gradually reduced in the same way as the shell plating. A minimum cross-sectional area of material alongside hatches must be maintained. Thus if part of the deck is cut away for a stairway or similar opening, compensation must be made in the form of either doubling plates or increased local plate thickness.

The deck forms a cover over the cargo, accommodation and machinery space and must therefore be watertight. The weather deck, and usually the second deck, are cambered to enable water to run down to the sides of the ship and hence overboard through the scuppers. The outboard deck strake is known as the *stringer plate* and at the weather deck is usually thicker than the remaining deck plating. It may be connected to the sheerstrake by means of a continuous *stringer angle* or *gunwale bar*.

Exposed steel decks above accommodation must be sheathed with wood which acts as heat and sound insulation. As an alternative the deck may be covered with a suitable composition. The deck must be adequately protected against corrosion between the steel and the wood or composition. The deck covering is stopped short of the sides of the deck to form a waterway to aid drainage.

Beams and Deck Girders

The decks may be supported either by transverse beams in conjunction with longitudinal girders or by longitudinal beams in conjunction with transverse girders.

The transverse beams are carried across the ship and bracketed to the side frames by means of *beam knees*. A continuous longitudinal girder is fitted on each side of the ship alongside the hatches. The beams are bracketed or lugged to the girders, thus reducing their span. With respect to the hatches, the beams are broken to allow open hatch space, and are joined at their inboard ends to either the girder or the hatch side coaming. A similar arrangement is necessary in way of the machinery casings. These broken beams are known as *half beams* and are usually shaped as bulb plates.

There are several forms of girder in use, two of which are shown in Figure 5.2.

If the girder is required to form part of the hatch coaming, the flanged girder (Figure 5.2a) is most useful since it is easy to produce and does not require the addition of a moulding to prevent chafing of ropes. Symmetrical girders such as in Figure 5.2(b) are more efficient but cannot form part of a hatch side coaming. Such girders must be fitted outboard of the hatch sides. The girders are bracketed to the transverse bulkheads and are supported at the hatch corners either by pillars or by hatch end

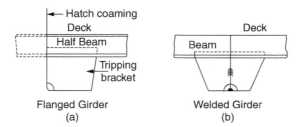

▲ **Figure 5.2** *Deck Girders*

girders extending right across the ship. Tubular pillars are most often used in cargo spaces since they give utmost economy of material and, at the same time, reduce cargo damage. In deep tanks, where hollow pillars should not be used, and in machinery spaces, either built pillars or broad flanged beams prove popular.

Most modern ships are fitted with longitudinal beams which extend, as far as practicable, along the whole length of the ship outside the line of the hatches. They are bracketed to the transverse bulkheads and are supported by transverse girders which are carried right across the ship, or, with respect to the hatches and machinery casings, from the side of the ship to the hatch or casing. The increase in continuous longitudinal material leads to a reduction in deck thickness. The portion of deck between the hatches may be supported either by longitudinal or transverse beams, neither having any effect on the longitudinal strength of the ship.

At points where concentrated loads are anticipated it is necessary to fit additional deck stiffening. Additional support is required with respect to winches, windlasses, deck cranes and capstans (Figure 5.3). The deck machinery is bolted to seatings which may be riveted or welded to the deck. The seatings are extended to distribute the load. With respect to the seatings, the beams are increased in strength by fitting reverse bars which extend to the adjacent girders. Solid pillars are fitted under the seatings to reduce vibration.

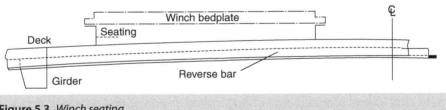

▲ **Figure 5.3** *Winch seating*

Hatches

Large hatches must be fitted in the decks of dry cargo ships to facilitate loading and discharging of cargo. It is usual to provide one hatch per hold or 'tween deck, although in ships having large holds two hatches are sometimes arranged. The length and width of hatches depend largely upon the size of the ship and the type of cargo likely to be carried. General cargo ships have hatches which will allow cargoes such as timber, cars, locomotives and crates of machinery to be loaded. A cargo tramp of about 10 000 tonne deadweight may have five hatches, each 10 m long and 7 m wide, although one hatch, usually to No. 2 hold, is often increased in length. Large hatches also allow easy handling of cargoes. Bulk carriers have long, wide hatches to allow the cargo to fill the extremities of the compartment without requiring trimming manually.

On container ships, the hatches will be designed to carry the weight of loaded containers. Here, several tiers of containers will be carried on top of the hatch covers. Designers will need to take note of the class rules and calculate the maximum permissible 'flexing' of the hatch cover. Calculations will be completed using Finite Element (FE) techniques. The calculations relating to the strength of the hatch coamings will also need to consider the maximum forces set up by the weight of the containers carried on top of the hatches, as well as the forces due to the pitching and rolling of the vessel.

The hatches are framed by means of hatch coamings which are vertical webs forming deep stiffeners. The heights of the coamings are governed by the Load Line Rules. On weather decks they must be at least 600 mm in height at the fore end and either 450 mm or 600 mm aft depending upon the draught of the ship. Inside superstructures and on lower decks no particular height of coaming is specified. It is necessary, however, for safety considerations, to fit some form of rail around any deck opening to a height of 800 mm. It is usual, therefore, at the weather deck, to extend the coaming to a height of 800 mm. In the superstructures and on lower decks portable stanchions are provided, the rail being in the form of a wire rope. These rails are only erected when the hatch is opened.

The weather deck hatch coamings must be 11 mm thick and must be stiffened by a moulding at the top edge. Where the height of the coaming is 600 mm or more, a horizontal bulb angle or bulb plate is fitted to stiffen the coaming which has additional support in the form of stays fitted at intervals of 3 m. Figure 5.4 gives a typical section through the side coaming of a weather deck hatch. The edge stiffening is in the form of a bulb angle set back from the line of the coaming. This forms a rest to support the portable beams. The edge stiffening on the hatch end coaming (Figure 5.5) used to be a Tyzack moulding which was designed to carry the ends of the wood boards.

The hatch coamings inside the superstructures are formed by 230 mm bulb angles or bulb plates at the sides and ends. The side coamings are usually set back from the opening to form a beam rest, while an angle is fitted at the ends to form a rest bar for the ends of the wood covers.

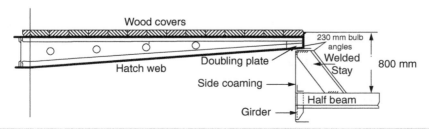

▲ **Figure 5.4** *Transverse section through hatch*

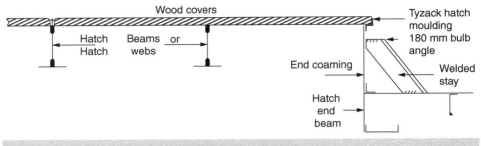

▲ **Figure 5.5** *Longitudinal section through hatch*

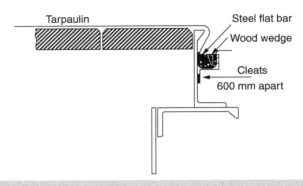

▲ **Figure 5.6** *Hatch closing arrangement*

The hatches used to be closed by wood boards which are supported by the portable hatch beams. The beams may be fitted in guides attached to the coamings and lifted out to clear the hatch, or fitted with rollers allowing them to be pushed to the hatch ends. The covers are made weathertight by means of tarpaulins which are wedged tight at the sides and ends (Figure 5.6), at least two tarpaulins being fitted on weather deck hatches.

Modern ships are fitted with steel hatch covers (Figure 5.7). There are many types available, from small pontoons supported by portable beams to the larger self-supporting type, the latter being the most popular. The covers are arranged in 4 to 6 sections extending right across the hatch and having rollers which rest on a runway. They are opened by rolling them to the end of the hatch where they tip automatically into the vertical position. The separate sections are joined by means of wire rope, allowing opening or closing to be a continuous action, a winch being used for the purpose.

Many other systems are available, some with electric or hydraulic motors driving sprocket wheels, some in which the whole cover wraps round a powered drum, while others have hydraulic cylinders built into the covers. In the latter arrangement pairs of covers are hinged together, the pairs being linked to provide continuity. Each pair of covers has one or two hydraulic rams which turn the hinge through 180°. The rams are actuated by an external power source, with a control panel on the side of the hatch coaming.

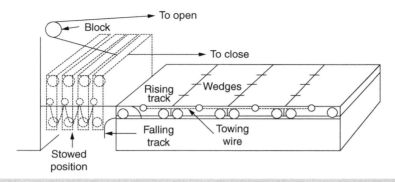

▲ **Figure 5.7** *Steel hatch covers*

The covers interlock at their ends and are fitted with packing to ensure that when the covers are wedged down, watertight cover is provided (Figure 5.8). Such covers do not require tarpaulins. At the hatch sides the covers are held down by cleats which may be manual as shown in Figure 5.9 or hydraulically operated.

Composite hatch covers have just been approved and fitted to the first bulk carrier during 2015. One significant problem has been that the existing rules were set around steel hatches. Therefore, none of the elements of the rule book fitted to the composite structure; this is however a logical development as the material is strong and light.

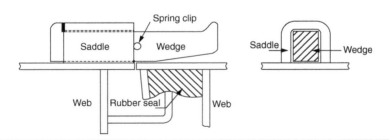

▲ **Figure 5.8** *Hatch wedges*

Hatch covers are very important and need to have the strength to withstand a pounding from the waves that could come over the side of the ship during heavy weather. 'Tween deck covers can also constructed of composite material, which makes them lightweight and easier to handle.

Deep tank hatches have two functions to fulfil (Figure 5.10). They must be watertight or oil-tight and thus capable of withstanding a head of liquid, and they must be large enough to allow normal cargoes to be loaded and discharged if the deep tank is required to act as a dry cargo hold. Such hatches may be 3 m or 4 m square. Because of

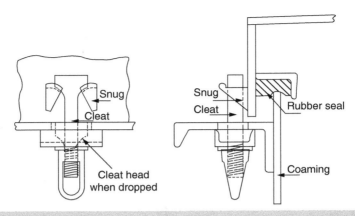

▲ **Figure 5.9** *Securing cleats*

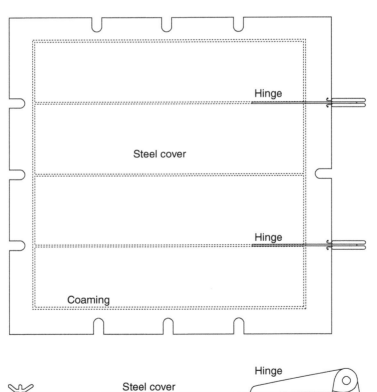

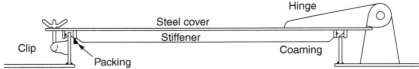

▲ **Figure 5.10** *Deep tank hatch*

the possible liquid pressure, the covers must be stiffened, while some suitable packing must be fitted in the coamings to ensure watertightness, together with some means of securing the cover. The covers may be hinged or arranged to slide.

Figure 5.11 summarises much of the information within this chapter by showing the relationship between the separate parts in a modern ship. The sizes or scantlings of the structure are typical for a ship of about 10,000 tonne deadweight.

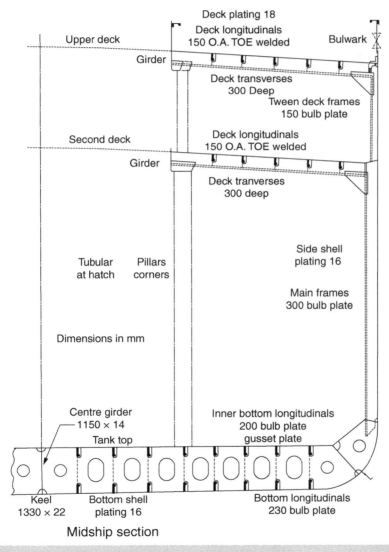

▲ **Figure 5.11** *Midship section of a welded ship*

6

BULKHEADS AND DEEP TANKS

There are three basic types of bulkheads used in ships; watertight bulkheads, tank bulkheads and non-watertight bulkheads. These bulkheads may be fitted longitudinally or transversely, although only non-watertight and some tank bulkheads are fitted longitudinally in most dry cargo ships. Their function is to keep enough buoyancy in the ship in the event of a rupture in the outer hull. This is a crucial concept and has been the ultimate aim of ship designers for many years. As we know from RMS *Titanic*, if enough watertight compartments are breeched, then the ship will sink.

However, as we also know from the *Titanic*, a watertight compartment not only has to be watertight, but it *must* also extend high enough in the vessel that it will not be breached, in order to maintain enough buoyancy to keep the vessel afloat in the event of a rupture of the hull (see Figure 6.1).

The International Maritime Organization's (IMO) Safe Return to Port (SRtP) for passenger ships, focuses on improving the survivability of the vessel in the event of a rupture of the hull. Designers must complete calculations based on compartments flooding that would not normally happen when the vessel is sailing routinely.

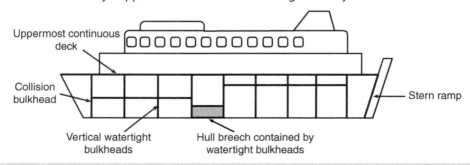

Uppermost continuous deck

Collision bulkhead

Stern ramp

Vertical watertight bulkheads

Hull breech contained by watertight bulkheads

▲ Figure 6.1

Watertight Bulkheads

The transverse watertight bulkheads of a ship have several functions to perform.

They divide the ship into watertight compartments and thus restrict the volume of water which may enter the ship if the shell plating is damaged. In passenger ships, complicated calculations are carried out to ensure an arrangement of bulkheads which will prevent the ship sinking even if it is damaged to a reasonable extent. A simpler form of calculation is occasionally carried out for cargo ships but results in only a slight indication of the likelihood of the vessel sinking, since the volume and type of cargo play an important role.

The IMO now requires that designers improve the 'survivability' of a vessel in the event of a collision with another ship or with a solid object. They must arrange the internal watertight structures accordingly. The watertight compartments also serve to separate different types of cargo and to divide tanks and machinery spaces from the cargo spaces. This new method of calculating the 'survivability' of a vessel, in the event of damage, is called the 'probabilistic' method of stability calculations (see page 174). The calculations work out the probability of a vessel surviving if the hull is holed in a specific position. The internal design of the vessel is then examined and changed, if necessary, to improve the 'survivability'.

In the event of fire, bulkheads significantly reduce the rate of spread of the fire. Controlling of the spread of fire, however, also depends upon the 'fire potential' on each side of the bulkhead, that is, the likelihood of the material near the bulkhead being ignited.

SOLAS Chapter 2 requires that vessels are constructed to significantly reduce the spread of fire. When a vessel is designed, this is achieved by dividing the vessel into different sections by using insulated bulkheads made of steel, or other material that is equivalent to steel. These are known as class 'A' or 'B' fire divisions.

Fire divisions made up of bulkheads and approved non-combustible material are further subdivided into 'A-60, A-30, A-15, A-0, B-15 and B-0. The number after the class standard A or B relates to the number of minutes that the bulkhead + insulation, is required to restrict the rise in temperature on the opposite side to a fire. For example, A-60 relates to the A standard for 60 minutes.

There is a class 'C' division, but this is only to be made of non-combustible material and is not subject to the rigorous testing of the class A and B divisions.

The class A division standard means that the temperature on the opposite side to the fire must be restricted to a rise of not more than 140°C, and a peak temperature of 180°C above the initial temperature. The length of time for this restriction will relate to the number of minutes specified in its classification outlined on page 97 (for further information please see page 177).

The transverse strength of the ship is also increased by the bulkheads which have much the same effect as the ends of a box. They prevent undue distortion of the side shell and reduce racking considerably.

Longitudinal deck girders and deck longitudinals are supported at the bulkheads which therefore act as pillars, while at the same time they tie together the deck and tank top and hence reduce vertical deflection when the compartments are full of cargo. This whole structure contributes to the ultimate strength of the hull girder.

Thus it appears that the shipbuilder has a very complicated structure to design. In practice, however, it is found that a bulkhead required to withstand a load of water in the event of flooding will readily perform the remaining functions.

The number of bulkheads in a ship depends upon the length of the ship and the position of the machinery space. Each ship must have a collision bulkhead and from 2010 this should be at least *0.05L* or 10 m (whichever is the least) from the forward perpendicular and no greater than *0.08L* or *0.05L* + 3 m (whichever is greater). The bulkhead must be continuous up to the uppermost continuous deck (also called the 'bulkhead deck', see page 32). Special considerations are made for Ro-Ro vessels that have a bow door.

The stern tube must be enclosed in a watertight compartment formed by the sternframe and the after peak bulkhead which may terminate at the first watertight deck above the waterline. A bulkhead must be fitted at each end of the machinery space although, if the engines are aft, the after peak may form the after boundary of the space.

In certain ships this may result in the saving of one bulkhead. In ships more than 90 m in length, additional bulkheads are required, with the actual number depending upon the length. Thus, a ship 140 m long will require a total of 7 bulkheads if the machinery is amidships or 6 bulkheads if the machinery is aft, while a ship 180 m in length will require 9 or 8 bulkheads, respectively. These bulkheads must extend to the freeboard deck and should preferably be equally spaced in the ship. However, from Chapter 1 of this book, students will realise that the holds in a ship may not be of equal length. The bulkheads are fitted in separate sections between the tank top and the lowest deck, and in the 'tween decks.

In ships employing diesel electric propulsion systems the actual engines might not be in line across the vessel. Therefore, to keep the propulsion systems separate, giving improved redundancy, additional watertight transverse and longitudinal bulkheads would be required.

Watertight bulkheads are formed by plates which are attached to the shell, deck and tank top by welding (Figure 6.2). Since water pressure increases with the head of water, and the bulkhead is to be designed to withstand such a force, it may be expected that the plating on the lower part of the bulkhead is thicker than that at the top. The bulkheads are supported by vertical stiffeners spaced 760 mm apart. Any variation in this spacing results in variations in size of stiffeners and thickness of plating. The ends of the stiffeners are usually bracketed to the tank top and deck although in some cases the brackets are omitted, resulting in heavier stiffeners.

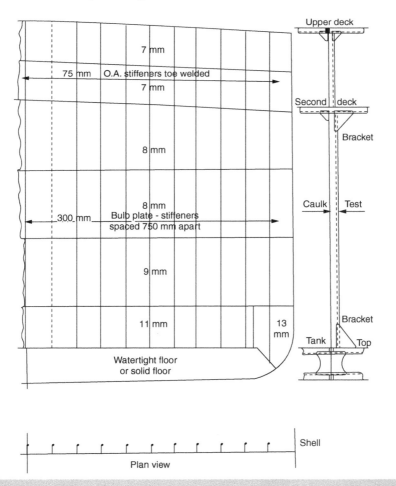

▲ **Figure 6.2** *Welded watertight bulkhead*

The stiffeners are in the form of either bulb plates or toe welded angles. It is of interest to note that since a welded bulkhead is less liable to leak under load, or alternatively it may deflect further without leakage, the strength of the stiffeners may be reduced by 15%. It may be necessary to increase the strength of a stiffener which is attached to a longitudinal deck girder in order to carry the pillar load.

The May 2010 revision of IACS rules requires that bulkheads are tested for watertightness by the application of a hydrostatic test. If this is not possible, then hydropneumatic testing can be used. This is where a tank is part filled with water and then pressurised with the use of compressed air. If neither of these methods are feasible, then they can be tested using water pressure of 200 kN/m^2 from a hose with a nozzle of at least 12 mm diameter applied from a distance of no more than 1.5 metres.

The hose test is carried out from the side on which the stiffeners are attached. It is essential that the structure should be maintained in a watertight condition. If it is found necessary to penetrate the bulkhead, precautions must be taken to ensure that the bulkhead remains watertight. If the after engine room bulkhead is penetrated by the main shaft, which passes through a watertight gland, and by an opening leading to the shaft tunnel, then this opening must be fitted with a sliding watertight door.

When pipes or electric cables pass through a bulkhead, the integrity of the bulkhead must be maintained. Figure 6.3 shows a bulkhead fitting in the form of a watertight gland for an electric cable.

In many insulated ships, ducts are fitted to provide efficient circulation of cooled air to the cargo spaces. The majority of such ships are designed so that the ducts from the hold spaces pass vertically through the deck into a fan room, separate rooms being constructed for each hold. In these ships it is not necessary to penetrate any transverse bulkhead with a duct. In some cases, however, it is necessary to penetrate the bulkhead in which case a sliding watertight shutter must be fitted.

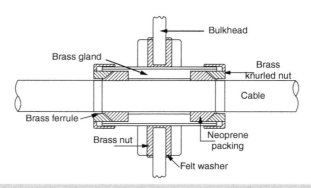

▲ **Figure 6.3** *Watertight cable gland*

Watertight Doors

A watertight door will be fitted to any access opening in a watertight bulkhead. Such openings must be cut only where necessary for the safe working of the ship and are kept as small as possible, 1.4 m high and 0.75 m wide being usual. The doors may be mild steel, cast steel or cast iron, and could be either vertical or horizontal sliding, the choice being usually related to the position of any fittings on the bulkhead. The latest class rules are that the doors should be strong enough to withstand the pressure of water that it could be subjected to. The method of construction would be a matter for the designer but on ships where the door is to be closed in operation at sea the door must be of the sliding type.

The means of closing the doors must be positive, that is, they must not rely on gravity or a dropping weight. The older type of vertical sliding doors (Figure 6.4) are closed by means of a vertical screw thread which turns in a gunmetal nut secured to the door.

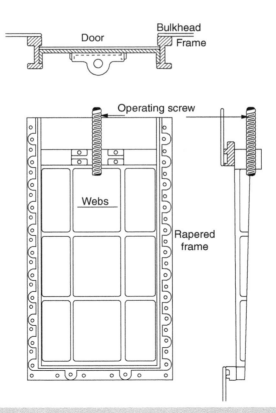

▲ **Figure 6.4** *Vertical sliding watertight door*

The screw is turned by a spindle which extends above the bulkhead deck, fitted with a crank handle allowing complete circular motion. A similar crank must be fitted at the door. The door runs in vertical grooves which are tapered towards the bottom, the door having similar taper, so that a tight bearing fit is obtained when the door is closed. Brass facing strips are fitted to both the door and the frame. There must be no groove at the bottom of the door to collect dirt which would prevent the door fully closing. An indicator must be fitted at the control position above the bulkhead deck, showing whether the door is open or closed.

A horizontal sliding door is shown in Figure 6.5. It could be operated by means of an electric motor 'A' which turns a vertical shaft 'B'. However, the more usual arrangement is by using a hydraulic system.

Near the top and bottom of the door, horizontal screw shafts 'C' are turned by the vertical shaft through the bevel gears 'D'. The door nut 'E' moves along the screw shaft within the nut box 'F' until any slack is taken up or the spring 'G' is fully compressed, after which the door moves along its wedge-shaped guides on rollers 'H'.

The door may be opened or closed manually at the bulkhead position by means of a handwheel J, the motor being automatically disengaged during this operation. An alarm bell gives a warning 10 s before the door is to close and while it is being closed. Opening and closing limit switches K are built into the system to prevent overloading of the motors.

A de-wedging device (Figure 6.6) may be fitted to release the door from the wedge frame and to avoid overloading the power unit if the door meets an obstruction. As the door-operating shaft turns, the spring-loaded nut E engages a lever L which comes into contact with a block M on the door frame. As the nut continues to move along the shaft, a force is exerted by the lever on the block, easing the door out of the wedge. Should a solid obstruction be met, the striker N lifts a switch bar P and cuts out the motor.

Modern door systems are usually hydraulically operated, having both remote and local operation, and they have a pumping plant which consists of two units. Each unit is capable of operating all the watertight doors and the electric motor is connected through the emergency power source. The doors may be closed at the door position or from a control point such as the bridge. If closed from the control point they may be opened from a local position, switches being fitted on both sides of the bulkhead, but close automatically when the switch is released. Alarms are required either side of the door locally and will be activated if the remote operation is started.

A electric motor
B vertical shaft
C horizontal shafts
D bevel gears
E door nut
F nut box
G spring

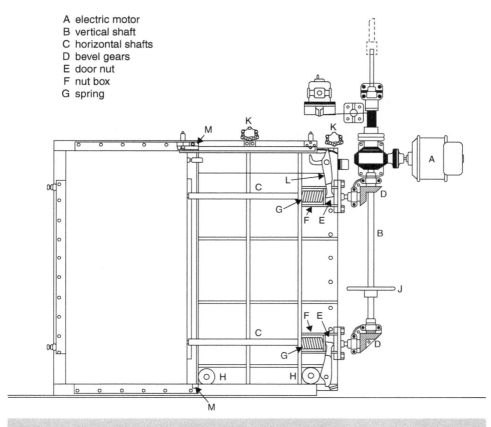

▲ **Figure 6.5** *Horizontal sliding watertight door*

H rollers
J handwheel
K limit switches
L de-wedging lever
M de-wedging block
N striker
P swtich bar

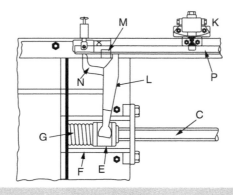

▲ **Figure 6.6** *De-wedging device*

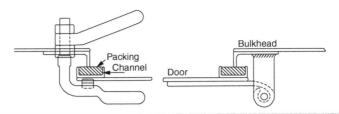

▲ **Figure 6.7** *Clip and hinge watertight door*

Watertight doors for all ships can be tested before fitting by a hydraulic pressure equivalent to a head of water from the door to the bulkhead deck. All such doors are hose tested after fitting.

Hinged watertight doors may be fitted to watertight bulkheads in passenger ships, above decks which are 2.2 m or more above the load waterline. Similar doors are fitted in cargo ships to weather deck openings which are required to be watertight. The doors are secured by clips which may be fitted to the door or to the frame. The clips are forced against brass wedges. The hinges must be fitted with gunmetal pins. Some suitable packing is fitted round the door to ensure that it is watertight. Figure 6.7 shows the hinge and clip for a hinged door, six clips being fitted to the frame.

Deep Tanks

It might be necessary, in ships with machinery amidships, to arrange a deep tank forward of the machinery space to provide sufficient ballast capacity in order to help trim the vessel correctly. This deep tank might also be designed to allow dry cargo to be carried and some ships may carry vegetable oil or oil fuel as cargo; however no flammable liquid can be carried in a deep tank forward of the collision bulkhead.

Deep tanks may also be provided for the carriage of oil fuel to be used as a ship's bunker fuel. The structure in these tanks is designed to withstand a head of water up to the top of the overflow pipe, the tanks being tested to this head or to a height of 2.44 m above the top of the tank, whichever is greater. It follows, therefore, that the strength of the structure must be much superior to that required for dry cargo holds. Where this is intended the designers must make this clear in their plans.

If a ship is damaged with respect to a hold, the end bulkheads are required to withstand the load of water without serious leakage. Permanent deflection of the bulkhead may be accepted under these conditions and a high stress may be allowed. There must be no permanent deflection of a tank bulkhead, however, and the allowable stress in the stiffeners must therefore be much smaller. The stiffener spacing on the transverse bulkheads is usually about 600 mm and the stiffeners are much heavier than those on hold bulkheads. If, however, a horizontal girder is fitted on the bulkhead, the size of the stiffeners may be considerably reduced. The ends of the stiffeners are bracketed, the toe of the bottom bracket being supported by a solid floor plate. The thickness of bulkhead plating is greater than required for hold bulkheads, with a minimum thickness of 7.5 mm. The arrangement of the structure depends upon the use to which the tank will be put.

Deep tanks for water ballast or dry cargo only

A water ballast tank should be either completely full or empty while at sea and therefore there should be no movement of water. The side frames are increased in strength by 15% unless horizontal stringers are fitted, when the frames are reduced. If such stringers are fitted, they must be continued across bulkheads to form a ring. These girders are substantial, with stiffened edges. The deck forming the top of a deep tank may be required to be increased in thickness because of the increased load due to water pressure. The beams and deck girders with respect to a deep tank are calculated in the same way as the bulkhead stiffeners and girders and therefore depend upon the head to which they are subject.

Deep tanks for oil fuel, oil cargo or fresh water

A partially full deep tank carrying oil or water will have a free surface and will therefore be subjected to dynamic forces. In the case of an oil fuel or fresh water bunker tank they will also be subjected to different levels of forces during the voyage. This results in reduced stability, while at the same time the momentum of the liquid moving across the tank may cause damage to the structure. To reduce this surging it is necessary to fit divisions or deep swashes to minimise the dynamic stresses caused by this arrangement.

These divisions may be intact, in which case they must be as strong as the boundary bulkheads, or perforated, when the stiffeners may be considerably reduced. The perforations must be between 5% and 10% of the area of the bulkhead. Any smaller area would allow a build-up of pressure on one side, for which the bulkhead is not designed, while a greater area would not reduce surging to any marked extent.

Sparring must be fitted to the cargo side of a bulkhead which is a partition between a bunker and a hold. If a settling tank is heated and is adjacent to a compartment which may carry coal or cargo, the structure outside the tank must be insulated. Figure 6.8 shows the structural arrangement of a deep tank which may be used for oil or dry cargo.

If dry cargo is to be carried in a deep tank, one or two large watertight hatches are required in the deck as described in Chapter 5.

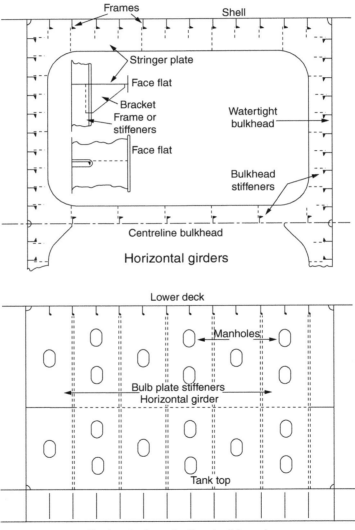

▲ **Figure 6.8** *Horizontal girders and perforated centreline bulkhead*

Non-watertight Bulkheads

Any bulkhead which does not form part of a tank or part of the watertight subdivision of the ship may be non-watertight. Many of these bulkheads are fitted in a ship, forming engine casings and partitions in accommodation. 'Tween deck bulkheads fitted above the freeboard deck may be of non-watertight construction, while many ships are fitted with partial centreline bulkheads if grain is to be carried. Centreline bulkheads and many deck-house bulkheads act as pillars supporting beams and deck girders, in which case the stiffeners are designed to carry the load. The remaining bulkheads are lightly stiffened by angle bars or welded flats.

Corrugated Bulkheads

A corrugated plate is stronger than a flat plate if subjected to a bending moment or pillar load along the corrugations. This principle may be used in bulkhead construction, when the corrugations may be used to dispense with the stiffeners (Figure 6.9), resulting in a considerable saving in weight. The troughs are vertical on transverse bulkheads but must be horizontal on continuous longitudinal bulkheads which form part of the longitudinal hull girder strength of the ship. The bulkhead must be provided with adequate support to avoid stress concentrations. They should be supported by floors or girders with the stiffening members providing additional support if necessary.

A load acting across the corrugations will tend to cause the bulkheads to fold in concertina fashion. It is usual, therefore, on transverse bulkheads to fit a stiffened flat plate at the shell, thus increasing the transverse strength. This method also simplifies the fitting of the bulkhead to the shell which may prove difficult where the curvature of the shell is considerable. Horizontal diaphragm plates are fitted to prevent collapse of the troughs. These bulkheads form very smooth surfaces which, in oil tanks, allows improved drainage and ease of cleaning. A vertical stiffener is usually necessary if the bulkhead is required to support a deck girder.

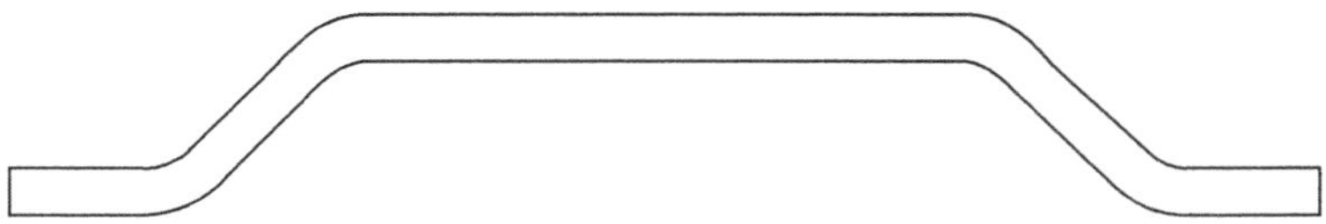

▲ **Figure 6.9** *Corrugated bulkhead*

7

FORE END
ARRANGEMENTS

The structural arrangements of different ships vary considerably, therefore typical and general examples are given in this section (see Figure 7.1).

Stem

The stem is formed by a solid bar which runs from the keel to the loaded waterline at the very front of the vessel's bow. The shell plating is stopped about 10 mm from the fore edge of the bar in order to protect the plate edges. At the bottom, the foremost keel plate is wrapped round the bar and is known as a *coffin plate*, due to its shape. A similar form of construction is used at the top of the stem.

The 'stem' bar is solid and rounded which improves the appearance considerably, particularly where the keel and side plates overlap.

Above the stem bar the stem is formed by plating which is strengthened by a welded stiffener on the centreline, the plating being thicker than the normal shell plating near the waterline but reduced in thickness towards the top. The plate stem is supported at intervals of about 1.5 m by horizontal plates known as *breast hooks*, which extend from the stem to the adjacent transverse frame. The breast hooks are welded to the stem plate and shell plating and are flanged on their free edge.

Modern stems are raked at 15°–25° to the vertical, with a large curve at the bottom, running into the line of the keel. Above the waterline some stems curve forward of the normal rake line to form a *clipper* bow. The flare of the bow is an interesting feature as it increases the water plane area the deeper it is immersed in the water due to the

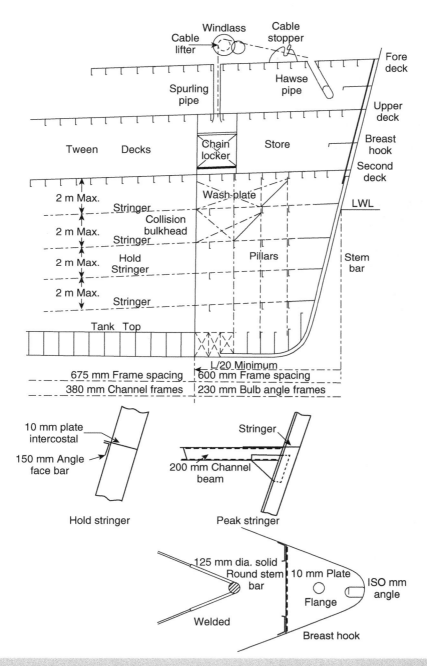

▲ **Figure 7.1** *Fore end construction*

pitching motion of the ship. This increase in area in turn increases the ship's buoyancy, thus helping to resist the pitching, and the additional forces must be taken into account when calculating the maximum bending moment caused by wave action. The downside of this arrangement is the added possibility of damage due to pounding.

Studies into a condition known as 'parametric rolling' have found that it comes about due to this fluctuating change in the transverse stability which in turn is encouraged by the changes in the waterplane area due to the vessel pitching.

Arrangements to Resist Panting

The structure of the ship is strengthened to resist the effects of panting from 15% of the ship's length from forward to the stem and aft of the after peak bulkhead.

In the fore peak, side or panting stringers are fitted to the shell at intervals of 2 m below the lowest deck (Figure 7.2). No edge stiffening is required as long as the stringer is connected to the shell, a welded connection being used in modern ships. Depending upon the size of the vessel, and the class rules, panting stringers may be omitted if the

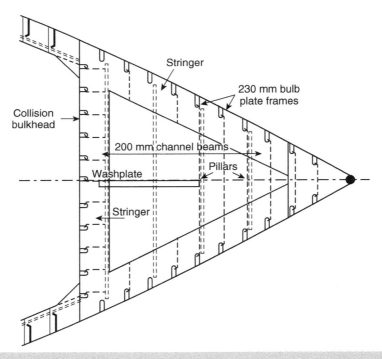

▲ **Figure 7.2** *Panting stringer*

hull thickness in increased. The side stringers meet at the fore end, while in many ships a horizontal stringer is fitted to the collision bulkhead in line with each shell stringer. This forms a ring round the fore peak tank and supports the bulkhead stiffeners. Channel beams are fitted at alternate frames in line with the stringers, and connected to the frames by brackets. The intermediate frames are bracketed to the stringer. The free edge of the bulkhead stringer may be stiffened by one of the beams. In fine ships it is common practice to plate over the beams, lightening holes being punched in the plate.

The tank top is not carried into the peak, but solid floors are fitted at each frame. These floors are slightly thicker than those in the double bottom space and are flanged on their free edge.

An interesting development is the X-bow (see Figure 7.3) design that was originally designed by the Norwegian shipbuilders 'Ulstein' for use on offshore vessels. The design gives the vessel an increasing underwater volume as the vessel pitches, similar to the flare in the bow of a conventional ship. This reduces the motion of the vessel due to wave action. The design is now finding its way on to passenger ships where the comfort of the passengers, and crew, is a prime concern. The same consideration can be extended to the comfort of the crew working on conventional ships that are operating in rough seas for long periods.

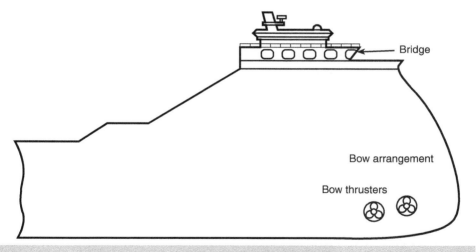

▲ **Figure 7.3** *X-bow type design for offshore vessels*

In the traditional arrangement, side frames are spaced 610 mm apart and, being so well supported, are much smaller than the normal hold frames. The deck beams are supported by vertical angle pillars on alternate frames, which are connected to the panting beams and lapped onto the solid floors. A partial wash-plate is usually fitted to reduce the movement of the water in the tank. Intercostal plates are fitted for 2 or 3 frame spaces in line with the centre girder. The lower part of the peak is usually filled with cement to ensure efficient drainage of the space.

Between the collision bulkhead and 15% length from forward, the main frames, together with their attachment to the margin, are increased in strength by 20%. In addition, the spacing of the frames from the collision bulkhead to 20% of the length from forward must be 700 mm. Light side stringers are fitted in the panting area in line with those in the peak. These stringers consist of intercostal plates connected to the shell and to a continuous face angle running along the toes of the frames. These stringers may be dispensed with if the shell plating is increased in thickness by 15%. This proves uneconomical when considering the weight but reduces the obstructions to cargo stowage in the hold. The peak is usually used as a tank and therefore such obstructions are of no importance.

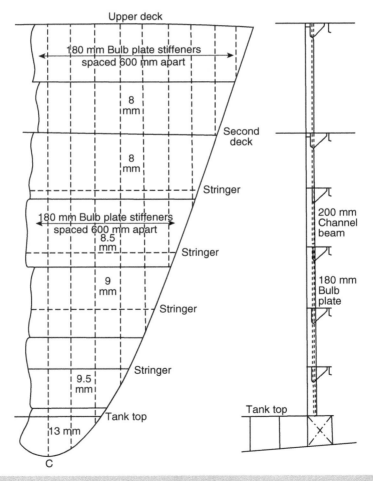

▲ **Figure 7.4** *Collision bulkhead*

The collision bulkhead is stiffened by vertical bulb plates spaced about 600 mm apart inside the peak. It is usual to fit horizontal plating because of the excessive taper on the plates which would occur with vertical plating. Figure 7.4 shows the construction of a collision bulkhead.

The structure in the after peak is similar in principle to that in the fore peak, although the stringers and beams may be fitted 2.5 m apart. The floors should extend above the stern tube or the frames above the tube must be stiffened by flanged tie plates to reduce the possibility of vibration. The latter arrangement is shown in Figure 8.8 in Chapter 8.

It is recommended that the bow of an ice breaking vessel is constructed without a bulbous bow. However, a vessel with limited ice breaking capacity should be constructed with added strength as detailed in the new 'polar code' that has recently been developed by the International Maritime Organization (IMO).

Arrangements to Resist Pounding

The structure is strengthened to resist the effects of pounding from the collision bulkhead to 25% of the ship's length from forward. The flat bottom shell plating adjacent to the keel on each side of the ship is increased in thickness by between 15% and 30% depending upon the length of the ship, larger ships having smaller increases.

In addition to increasing the plating, the unsupported panels of plating are reduced in size. In transversely framed ships the frame spacing in this region is 700 mm compared with 750–900 mm amidships. Longitudinal girders are fitted 2.2 m apart, extending vertically from the shell to the tank top, while intermediate half-height girders are fitted to the shell, reducing the unsupported width to 1.1 m. Solid floors are fitted at every frame space and are attached to the bottom shell by continuous welding.

If the bottom shell of a ship is longitudinally framed, the spacing of the longitudinals is reduced to 700 mm and they are continued as far forward as practicable to the collision bulkhead. The transverse floors may be fitted at alternate frames with this arrangement

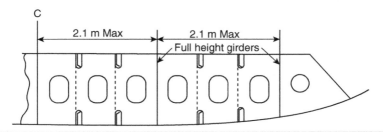

▲ **Figure 7.5** *Solid floor in pounding area*

and the full-height side girders may be fitted 2.1 m apart. Half-height girders are not required (Figure 7.5).

Bulbous Bow

One of the most important tasks facing designers of ships is to optimise hull performance. There are two aspects to this activity: the first is to reduce hull resistance and the second is to reduce the effect of 'wave drag'.

The act of pushing the vessel through the water will create a bow wave. If a sphere is immersed just below the surface of the water and at the bow of the ship the wave from the sphere interferes with the normal bow wave created by the vessel and results in a smaller bow wave. Thus the force required to produce the bow wave is reduced.

At the same time, however, the wetted surface area of the ship is increased, causing a slight increase in the frictional resistance. In slow ships the effect of a bulbous bow could be an increase in the total resistance, but in faster ships, where the wave making resistance forms a large proportion of the total resistance, the latter is reduced by fitting a bulbous bow. A bulbous bow also increases the buoyancy forward and hence reduces the pitching of the ship to some small degree. Optimising the design of the bulbous bow is very important. On modern ships the operating range might also be increased and therefore the design of the bow might not be as straight forward as first thought by the designers.

Although the actual design of a bulbous bow will be optimised to an actual vessel, a typical construction of the bulbous bow is shown in Figure 7.6. The stem plating is formed by steel plates supported by a centreline web and horizontal diaphragm plates 1 m apart. The outer bulb plating is thicker than the normal shell plating, partly because of high water pressures and partly due to the possible damage by anchors and cables.

It is often found that due to the reduced width at the waterline caused by the bulb, horizontal stringers in the fore peak prove uneconomical and complete perforated flats are fitted. The framing of the bulb could be joined to the general frame of the fore peak using diaphragm plates.

Anchor and Cable Arrangements

A typical arrangement for raising, lowering and stowing the anchors of a ship is shown in Figure 7.1. The anchor is attached to a heavy chain cable which is led through the *hawse pipe* over the windlass and down through a *chain pipe* or *spurling pipe* into the chain locker.

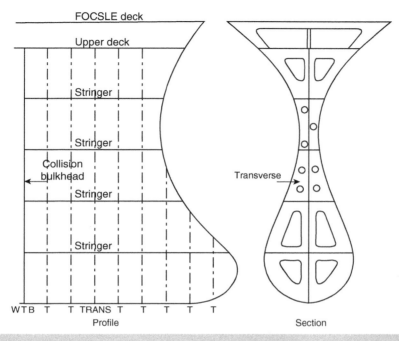

▲ **Figure 7.6** *Bulbous bow*

The hawse pipes may be constructed of mild steel tubes with castings at the deck and shell, or cast in one complete unit for each side of the ship. There must be ample clearance for the anchor stock to prevent jamming and they must be strong enough to withstand the hammering which they receive from the cable and the anchor. The shell plating is increased in thickness with respect to each hawse pipe and adjacent plate edges are fitted with mouldings to prevent damage. A chafing piece is fitted to the top of each hawse pipe, while a sliding cover is arranged to guard the opening.

The *cable stopper* is a casting with a hinged lever, which may be used to lock the cable in any desired position and thus relieve the load from the windlass either when the anchor is out or when it is stowed.

The drums of the windlass are shaped to suit the cable and are known as *cable lifters*. The cable lifters are arranged over the spurling pipes to ensure a direct lead for the cables into the lockers. The windlass may be either steam or electric in common with the other deck auxiliaries. Warping ends are fitted to assist in handling the mooring ropes. The windlass must rest on solid supports with pillars and runners with respect to the holding down bolts, a 75 mm teak bed being fitted directly beneath the windlass.

The chain pipes are of mild steel, bell mouthed at the bottom. The bells may be of cast iron, well rounded to avoid chafing. The pipes are fitted as near as possible to the centre of the chain locker for ease of stowage.

The chain locker may be fitted between the upper and second decks, below the second deck or in the forecastle. It must be of sufficient volume to allow adequate headroom when the anchors are in the stowed position. The locker is usually situated forward of the collision bulkhead, using this bulkhead as the after locker bulkhead. The locker is not normally carried out to the ship side. The stiffeners are preferably fitted outside the locker to prevent damage from the chains. If the locker is fitted in the forecastle, the bulkheads may be used to support the windlass. A centreline division is fitted to separate the two chains and is carried above the stowed level of the chain but is not taken up to the deck. It is stiffened by means of solid half-round bars while the top edge is protected by a split pipe. Foot holds are cut in to allow access from one side to the other. A hinged door is fitted in the forward bulkhead, giving access to the locker from the store space. Many lockers are fitted with false floors to allow drainage of water and mud, which is cleared by a drain plug in the forward bulkhead, leading into a drain hat from where it is discharged by means of a hand pump. The end of the cable must be connected to the deck or bulkhead in the chain locker. A typical arrangement is shown in Figure 7.7.

With this method, use is made of the existing stiffeners fitted to the fore side of the collision bulkhead. Two similar sections are fitted horizontally back to back, riveted to the bulkhead and welded to the adjacent stiffeners. A space is allowed between the horizontal bars to allow the end link of the cable to slide in and be secured by a bolt.

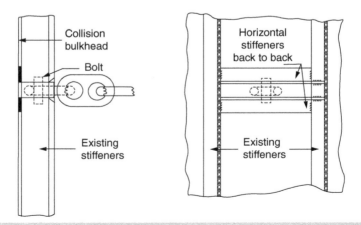

▲ **Figure 7.7** *Connection of end of chain*

AFTER END ARRANGEMENTS

Continuing with the task of optimising hull performance, the after end arrangement has been the subject of considerable focus in recent years. This is because so much of the aft end depends upon the requirements of the steering and propulsion machinery.

Traditionally ships have one or two propellers, being driven by shafts that need to project out through the hull via a seal, called the stern tube seal. The strength of the after end arrangement will be determined by the:

- weight of the propeller
- dynamic loading
- vibration analysis (including excessive conditions due to the loss of a blade).

Ships have also traditionally been turned by means of a rudder (see page 120). However, with the development of electric 'podded' drive motors (see page 136), the traditional requirements to accommodate the needs of the shaft line drive and the rudders for steering are no longer necessary.

This means that the designer can take a different and new way in optimising hull performance by considering the aft end arrangement that is much more efficient at handling the flow of water over the hull.

The two main basic arrangements are described in this chapter; however, the actual arrangements will differ depending upon the arrangement of the drive system.

Cruiser Stern

The cruiser stern forms a continuation of the hull of the ship above the sternframe and improves the appearance of the ship. It increases the buoyancy at the after end and improves the inflow of water to the propeller disk.

This arrangement is however very susceptible to slamming and must therefore be heavily stiffened. Class rules require that the framing should be similar to that of the after peak with web frames fitted where necessary, while solid floors must be fitted, together with a centreline girder. In practice the structure is arranged in two main forms:

1. Cant frames in conjunction with cant beams.
2. Horizontal frames in conjunction with either cant beams or transverse beams.

The description *cant frame* is rarely used by most classification societies but it is a frame that is set at an angle to the centreline of the ship. Such frames are fitted 610 mm apart, thus dividing the perimeter of the cruiser stern into small panels. At the top, these frames are bracketed to *cant beams* which also lie at an angle to the centreline. The forward ends of the cant beams are connected to a deep beam extending right across the ship. At the lower ends, the cant frames are connected to a solid floor (Figure 8.1).

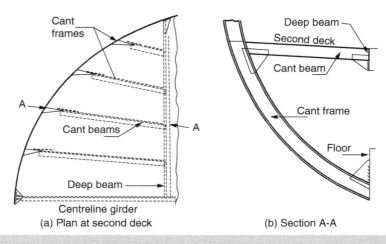

(a) Plan at second deck

(b) Section A-A

▲ **Figure 8.1** *(a) Plan at second deck and (b) section A-A*

The alternative method of construction has proved very successful, particularly with prefabricated structures. The horizontal frames are fitted at intervals of about 750 mm and are connected at their forward ends to a heavier transverse frame. They are supported at the centreline by a deep web which is also required with the cant frame system. If cant beams are fitted, the end brackets are carried down to the adjacent horizontal frame.

The structure at the fore end of the cruiser stern consists of solid floors attached to vertical side frames with transverse beams extending across the decks. A watertight rudder trunk is fitted enclosing part of the rudder stock. A typical centreline view of a cruiser stern with horizontal framing is shown in Figure 8.2.

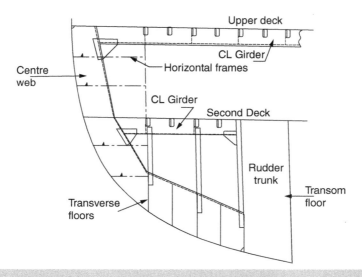

▲ **Figure 8.2** *Cruiser stern*

Many larger ships, especially passenger ships, are constructed with a transom stern. In this arrangement the stern is flat (Figure 8.3). The transom stern construction is more cost-effective, at the same time reducing the bending moment on the after structure caused by the unsupported overhang of the cruiser stern. The stiffening is invariably horizontal. Where the podded drive is to be accommodated as in the *Queen Mary II*, a 'rounded' compromise to the transom stern has been used. The 'Costanzi' stern improves the sea keeping ability of the ship while retaining the need for a flat underside to fit the podded drives.

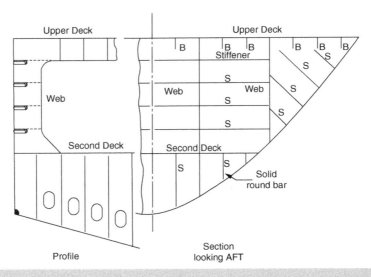

Upper Deck   Upper Deck

Web

Second Deck   Second Deck

Stiffener

Web

Web

Solid round bar

Profile

Section looking AFT

▲ **Figure 8.3** *Transom stern*

Sternframe and Rudder

The sternframe forms the termination of the lower part of the shell at the after end of the ship. The actual arrangement will be determined by the way that the propeller is sited and by the maximum allowable stresses and bending moments for the manufactured structure, as the aft end arrangement will have a considerable effect on the water flowing to the propeller. Cavitation and pressure fluctuations will be caused if this design is not correct. The classification societies set the criteria for the minimum clearance between the edge of the propeller and the hull of the vessel.

In single screw ships the sternframe carries the boss and supports the after end of the sterntube. The rudder is usually supported by a vertical post which forms part of the sternframe. It is essential, therefore, that the structure be soundly constructed and of tremendous strength. Sternframes may be cast, fabricated or forged, the latter method of construction having lost its popularity although parts of the fabricated sternframe may be forged. Both fabricated and cast sternframes may be shaped to suit the form of the hull and streamlined to reduce turbulence of the water. The choice of casting or fabrication in the construction of a sternframe depends upon the personal preference of the shipowner and shipbuilder, neither method having much advantage over the other.

It is useful to consider the rudder in conjunction with the sternframe. There are no fixed regulations for the area of a rudder, but it has been found in practice that an area of between one-sixtieth and one-seventieth of the product of the length and draught of the ship provides ample manoeuvrability for deep sea vessels. The ratio of depth to width, known as the aspect ratio, is usually about two. Single plate rudders were used for many years but are now seldom used because of the increased turbulence they create. Modern rudders are streamlined to reduce eddy resistance. If part of the rudder area lies forward of the turning axis, the turning moment is reduced and hence a smaller rudder stock may be fitted. A rudder with the whole of its area aft of the stock is said to be *unbalanced*. A rudder with between 20% and 40% of its area forward of the stock is said to be *balanced*, since at some rudder angle there will be no torque on the stock. A rudder which has part of its area forward of the stock, but at no rudder angle is balanced, is said to be *semi-balanced.*

Fabricated sternframe with unbalanced rudder

Figure 8.4 shows a fabricated sternframe used to support an unbalanced rudder. The *sole piece* is a forging which is carried aft to form the *lower gudgeon* supporting the bearing pintle, and forward to scarph to the aftermost keel plate. The sternpost is formed by a solid round bar to which heavy plates, 25–40 mm thick, are welded, the boss being positioned to suit the height of the shaft. Thick web plates are fitted horizontally to tie the two sides of the sternframe rigidly. The side shell plates are riveted or welded to the plates forming the sternframe. This form of construction is continued to form the *arch* which joins the sternpost to the rudderpost. Vertical webs are used in this position to secure the sternframe to the floor plates, while a thick centreline web is fitted to ensure rigidity of the arch.

The rudderpost is similar in construction to the sternpost, a thick web plate being fitted at the after side, while one or more gudgeons are fitted as required to support the rudder. The web plate is continued inside the ship at the top of the rudder post and attached to a thick transom floor which is watertight.

The rudder is formed by two plates about 10–20 mm thick, connected at the top and bottom to forgings which are extended to form the upper and lower gudgeons. The upper forging is opened into a palm, forming part of the horizontal coupling. This palm is stepped to provide a shoulder which reduces the possibility of shearing the bolts. The side plates are stiffened by means of vertical and horizontal webs but the structure is difficult to weld due to its inaccessibility. An efficient attachment may be made by fitting a flat bar to the edge of the horizontal webs and slot welding.

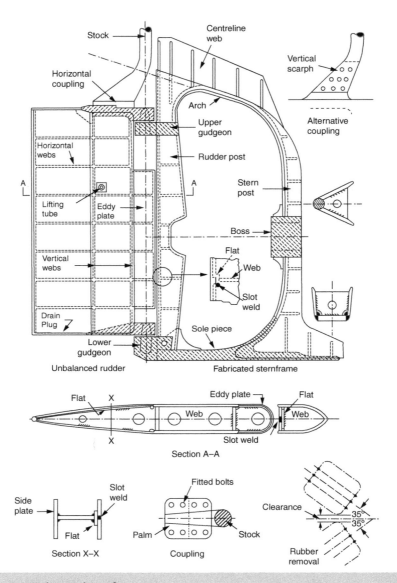

▲ **Figure 8.4** *Fabricated sternframe*

One disadvantage of double plate rudders is the possibility of internal corrosion and therefore the inner surfaces must be adequately protected by means of a suitable coating, while a drain plug should also be fitted to facilitate draining any water when dry-docking the vessel. It is however important that the inside to the rudder is kept dry to give the rudder buoyancy and reduce the loading on the carrier bearings.

The rudder is supported by *pintles* which fit into the gudgeons (Figure 8.5). The upper part of each pintle is tapered and fits into the rudder gudgeons. The pintle is pulled hard against the taper by means of a large nut with some suitable locking device, such as lock nut or split pin. A brass liner is fitted round the lower part of the pintle.

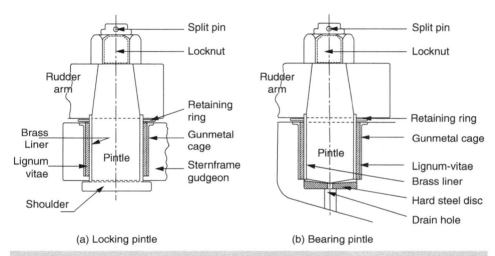

(a) Locking pintle

(b) Bearing pintle

▲ **Figure 8.5** *(a) Locking pintle and (b) bearing pintle*

Modern polymer based bearing material is dovetailed into the sternframe gudgeon to provide a bearing surface for the pintle, allowing it to turn but preventing any side movement. A head is fitted to the upper pintle to prevent undue vertical movement of the rudder. This is known as a *locking pintle*. The bottom pintle is known as a *bearing pintle* since it rests on a hardened steel pad shaped to suit the bottom of the pintle.

With the emphasis on environmentally friendly ships this is an important area for consideration, and the lubrication of the rudder bearings could mean that oil or grease has the potential to escape into the sea/river. Biodegradable lubricants should be used where self-lubricating polymer material is not being fitted.

The rudder is turned by means of a stock which is of forged steel, opened out into a palm at its lower end. The stock is carried through the rudder trunk and keyed to the steering engine. *It is essential that the centreline of stock and centreline of pintles are in the same line*, otherwise the rudder will not turn. A watertight gland must be fitted round the stock where it penetrates the deck. Many ships, however, are fitted with rudder

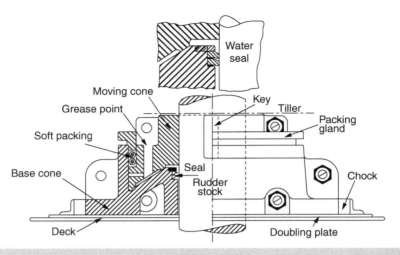

▲ **Figure 8.6** *Rudder carrier*

carriers (Figure 8.6), which themselves form watertight glands. The bearing surfaces are formed by cast iron cones, the upper cone being fitted to the rudder stock. As the bearing surfaces of the

lower pintle wear, the weight of the rudder will be taken by the carrier, and therefore the vertical wear down should be very small. Indeed, it is found in practice that any appreciable wear down is the result of a fault in the bearing surfaces, usually due to the misalignment of the stock. This causes uneven wasting of the surface and necessitates refacing the bearing surfaces and realigning the stock. In most cases, however, the cast iron work hardens and forms a very efficient bearing surface.

To remove the rudder it is first necessary to remove the locking pintle. The bearing may not be removed at this stage. The rudder is then turned by means of the stock to its maximum angle of, say, 35° on one side. The bolts in the coupling are removed and the stock raised sufficient to clear the shoulder on the palm. The stock is turned to the maximum angle on the opposite side, when the two parts of the coupling must be clear. The rudder may then be removed or the stock drawn from the ship.

Cast steel sternframe with balanced rudder

When a complex shape is required designers may specify a cast steel sternframe as shown in Figure 8.7. The casting may be in one or two pieces, the latter reducing the cost of repair in the event of damage. The sole piece is carried forward and scarphed to

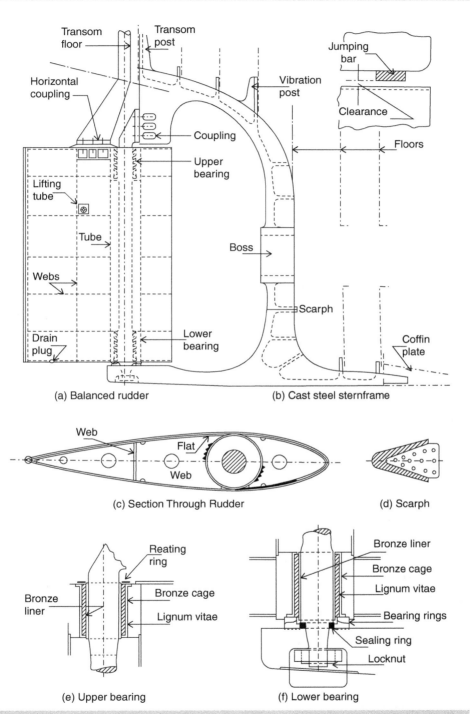

▲ **Figure 8.7** *(a) Balanced rudder; (b) Cast steel sternframe; (c) section through rudder; (d) scarph; (e) upper bearing; and (f) lower bearing*

the aftermost keel plate, while the after end forms the lower gudgeon. The sternpost is carried up inside the ship and opened to form a palm which is connected to a floor plate. This is known is a vibration post. The casting is continued aft at the top of the propeller aperture to form the arch and the upper rudder support. At the extreme after end the casting is carried inside the ship and opened into a palm which is connected to a watertight transom floor.

The rudder is constructed of double plates, with a large tube down the centre. The rudder post is formed by a detachable forged steel mainpiece which is carried through the tube, bolted to a palm on the stern frame at the top and pulled against a taper in the lower gudgeon. The main piece is increased in diameter at the top and bottom where suitable bearing strips are fitted. Castings are fitted at the top and bottom of the tube to carry the bearing strips and hard steel bearing rings are fitted between the rudder and the bottom gudgeon to take the weight of the rudder.

A horizontal coupling is shown in Figure 8.7a, attaching the stock to the rudder with the aid of fitted bolts. There must be sufficient vertical clearance between the stock and the mainpiece to allow the mainpiece to be raised sufficiently to clear the bottom gudgeon when removing the rudder. The upper stock is usually supported by a rudder carrier. By balancing a rudder in a particular ship, the diameter of the stock was reduced from 460 mm to 320 mm. This allows reduction in the thickness of the side plates and the size of the steering gear.

A flat bar is welded to the bottom of the horn to restrict the lift of the rudder. The clearance between the rudder and the flat should be less than the cross-head clearance. Any vertical force on the rudder will hence be transmitted to the sternframe and not to the steering gear.

Open water stern with spade rudder

Many modern large ships are fitted with spade-type rudders (Figure 8.8). The rudder is supported by means of a gudgeon on a large rudder horn and by the lower end of the stock, the latter being carried straight into the rudder and is either of a keyed or keyless design.

The lower part of the ship at the after end is known as the *deadwood* since it was seen in wooden ships to serve no useful purpose. However, as we now know the arrangement of the hull in this area can have a significant effect on hull performance.

The spade rudder is unsupported at the bottom and hence an open aperture is possible. This allows the deadwood to be cut away, resulting in a better flow of water to the propeller. In addition, the distance of the rudder from the propeller may be adjusted

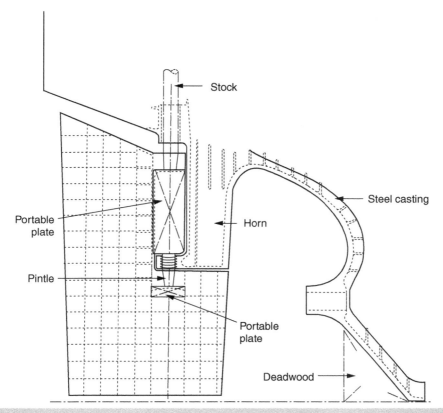

▲ Figure 8.8 *Spade rudder and open water stern*

to improve the efficiency of the rudder and in practice considerable reductions have been made, both in the diameter of the turning circle and in the vibration when turning. A recent design, from Rolls-Royce, extends the boss of the propeller to meet a similarly formed section on the rudder. This design drastically reduces the propeller boos vortices and improves fuel consumption.

The rudder horn and the sternframe may be cast or fabricated depending upon the complexity of the final shape required and the calculated stresses involved.

Rudder and sternframe for twin screw ship

In a twin screw ship the propellers are fitted off the centreline of the ship and therefore no aperture is required in the sternframe which may then be designed to only support the rudder. In a single screw ship, the 'deadwood' section is used to assist in the support of the sole piece but this is not necessary in twin screw vessels. Many designers make use of this space by carrying the lower part of the rudder forward of the centreline of stock. Figure 8.9 shows a typical arrangement.

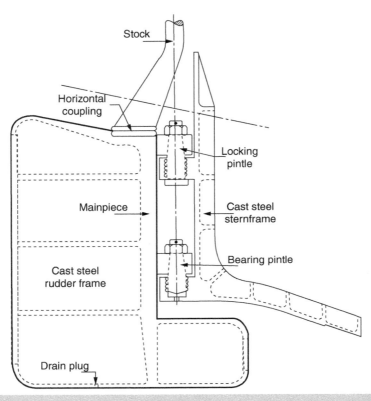

▲ **Figure 8.9** *Semi-balanced rudder – twin screw ship*

Where the sternframe is of cast steel and constructed to cut out the deadwood, it is notched or rebated to suit the shell plating. The top of the casting is connected by means of a palm to the transom floor.

The rudder shown in the diagram has a frame of cast carbon–manganese steel, the frame consisting of a mainpiece with horizontal arms which are of streamlined form. The side plates are riveted or welded to the frame. There must be sufficient clearance between the lower part of the sternframe and the extension of the rudder to allow the rudder to be lifted clear of the bearing pintle. The mainpiece must be particularly strong to prevent undue deflection of the lower, unsupported portion.

Bossings and spectacle frame for twin screw ship

The shafts of a twin screw ship are set at a small angle to the centreline. As the width of the ship reduces towards the after end, the shaft projects through the normal line of the shell. The shell widens out round the shaft to form the *bossings* which allow access to the shaft from inside the ship and allows bearings to be fitted where required. The after end of the bossing terminates in a casting which carries the boss and supports the after end of the stern tube. This casting must be strongly constructed and efficiently attached to the main hull structure to reduce vibration. It is usual to carry the casting in one piece across the ship, thus forming the *spectacle frame* (Figure 8.10). The centre of the casting is in the form of a box extending over two frame spaces and attached to thick floor plates.

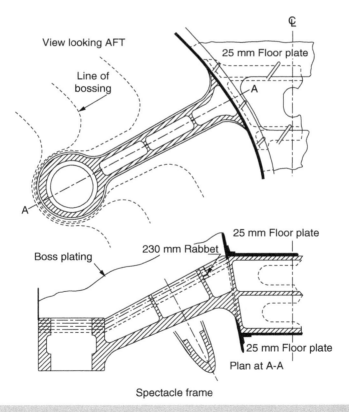

▲ **Figure 8.10** *View looking aft and spectacle frame*

Shaft Tunnel

When the machinery space is separated from the after peak by one or more cargo holds, the main shafting must be carried through the holds. A tunnel is then built round the shaft to prevent contact with the cargo and to give access to the shaft at all times for maintenance, inspection and repair. The tunnel is watertight and extends from the after machinery space bulkhead to the after peak bulkhead. It is not necessary to provide a passage on both sides of the shaft, and the tunnel is therefore built off the centreline of the ship, allowing a passage down the starboard side. The top of the tunnel is usually circular, except in a deep tank, when it is more convenient to fit a flat top. Figure 8.11 shows a cross-section through a shaft tunnel.

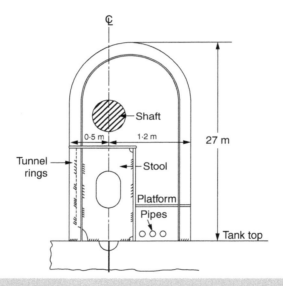

▲ **Figure 8.11** *Shaft tunnel*

The tunnel stiffeners or *rings* are fitted inside the tunnel although in insulated ships and in tunnels which pass through deep tanks, the rings are fitted outside the tunnel. The rings may be welded to the tank top or connected by angle lugs. The plating is attached to the tank top by welding or by a boundary angle fitted on the opposite side of the plating to the stiffeners. The stiffeners and plating must be strong enough to withstand a water pressure without appreciable leakage in the event of flooding. The scantlings are therefore equivalent to those required for watertight bulkheads. Under the hatches the tunnel top plating is increased by 2 mm unless wood sheathing is fitted. One of the side plates is arranged so that it may easily be removed, together with the stiffeners, to allow the main shafting to be unshipped.

A watertight door is fitted in the machinery space bulkhead giving access to the shaft tunnel from the machinery space (see Chapter 6). At the after end of the tunnel, a watertight escape trunk is fitted and extends to the deck above the load waterline. At the after end of the tunnel, the ship is so fine that there is very little useful cargo space at each side of the tunnel. The tunnel top is then carried right across the ship to form a *tunnel recess*. The additional space on the port side of this recess is usually used to store the spare tail shaft.

The shaft tunnel can be used as a pipe tunnel, the pipes being carried along the tank top with a light metal walking platform fitted about 0.5 m from the tank top. The shaft is supported at intervals by bearings which are fitted on shaft stools. The tops of the stools are lined up accurately to suit the height of the shaft, although adjustments to the height of bearings are made when the ship is afloat. The stools are constructed of 12 mm plates welded together. They are attached to the tunnel rings to prevent movement of the bearings which could lead to damage of the shaft. The loads from the bearings are transmitted to the double bottom structure by means of longitudinal brackets. Manholes are cut in the end plates to reduce the weight and to allow inspection and maintenance of the stools.

Thrust Improvement

With the cost of fuel being a significant operational expense when running a fleet of ships, anything that can be achieved to improve fuel efficiency is of great importance to the owner. It is also important for the industry and for society as a whole as better fuel efficiency means less damage to the environment.

Research into the causes of cavitation and propulsion induced vibration have produced some significant information. For example, just 40 years ago it was thought that vibration from the propeller was due to a resonating frequency that was due to a function of the number of blades and the propeller turning at certain revolutions.

Modern research shows that there are several forms of cavitation and uneven inflow of water to the propeller that are causing the vibration. If the circle scribed by the propeller is considered, students will be able to realise that the water flowing into the circle, or disc, at the lower section will not be as disturbed as the water flowing towards the propeller at the top of the disc. The reason for this is that the water at the top must come around the hull whereas the water at the bottom just runs straight across the bottom of the ship. This flow of water from around the hull is called the 'wake field' and a poor wake field gives rise to poor propeller efficiency (see Chapter 11 for more details).

This causes a pressure differential at different places around the propeller disc. The fluctuating pressure is happening each revolution (up to twice per second) and it is this pressure differential – together with the varying forms of cavitation – that is the designers focus when looking for efficiency gains.

One of the major developments in this area has been with the design of the 'podded drive'. This feature has enabled the designer to smooth out the inflow of water to the propeller disc and thus improve efficiency. Further gains are made as the pods are arranged to 'pull' rather than to 'push'.

The azimuthing ability also means that the designer can remove the rudder from the equation as the pods are also used to steer the ship. The next few pages contain information about some of the most significant 'thrust improvement' devices currently in use on ships.

Kort Nozzle

The Kort nozzle is a form of hollow truncated cone which is fitted around the propeller in order to increase the propulsive efficiency. Two types of nozzles are available: the fixed nozzle, which is welded to the ship and forms part of the hull structure; and the nozzle rudder, which replaces the normal rudder.

Fixed nozzle

Figure 8.12 shows the arrangement of a nozzle which is fixed and joined together by plates which form aerofoil cross-section. At the bottom the nozzle is welded to the sole piece of the sternframe, while at the top it is faired into the shell plating. Diaphragm plates are fitted at intervals to support the structure.

The nozzle directs the water into the propeller disc in lines parallel to the shaft, causing a possible increase in thrust of about 20–50%. This may be used in several ways. The ship speed may be increased by 5–10% with no increase in power. The power may be reduced for the same speed or, in vessels such as tugs, the increase in thrust or towing

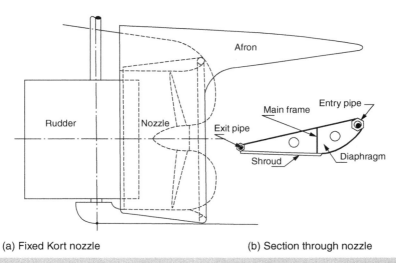

(a) Fixed Kort nozzle   (b) Section through nozzle

▲ **Figure 8.12** *(a) Fixed Kort nozzle and (b) section through nozzle*

force may be accepted for the same power. It is also found that, in rough water, the effect of pitching on the propulsive efficiency is greatly reduced, since the water is still directed into the propeller disc.

In all rotating machinery the clearance between the rotating element and the casing should preferably be constant, maximum efficiency being obtained when the clearance is in the order of one-thousandth of the diameter. The Kort nozzle permits a constant clearance, although this is in the order of one-hundredth of the diameter. One of the troubles with conventional type sternframes is the fluctuation in stress on the propeller blades as they pass close to the structure. If the tip clearance is too small, these fluctuations cause vibration of the propeller blades. Such vibration is avoided entirely with the constant tip clearance of the Kort nozzle.

The nozzle also acts as a guard for the propeller, protecting it from loose ropes and floating debris. If, however, ropes do become entangled with the propeller, they are much more difficult to remove than with the open propeller. Such nozzles are fitted mainly to tugs and trawlers, where the increased pull may be utilised directly in service.

Nozzle rudder

One of the disadvantages of the fixed nozzle was the difficulty of moving astern, the after end tending to drift. This is overcome by fitting a centreline fin plate to the after end of a nozzle which turns about a vertical stock and thus dispenses with the normal form of rudder (Figure 8.13). The water is then projected at an angle to the centreline, causing the ship to turn. The centreline of the stock must be in line with the propeller in order to allow the nozzle to turn and yet maintain small tip clearance.

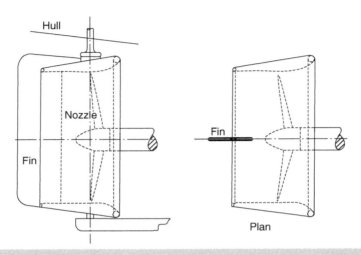

▲ **Figure 8.13** *Kort nozzle rudder*

The support of the shaft at the extreme after end may prove rather difficult with this form of construction. The bossing may be increased and extended, or the boss may be supported by brackets fitted to the stern.

Both types of nozzle may be fitted to existing ships, although the rudder type requires greater alteration to the structure.

Mewis duct

A reason for poor propeller performance was given earlier in this section as a reason for poor propeller performance. The Mewis duct is designed to smooth out the wake field thus reducing the pressure fluctuations around the propeller disc (Figure 8.14).

The duct is a similar shape to the Kort nozzle, but it is placed ahead of the propeller and acts a funnel for the water flowing into the propeller. The flow is further 'improved' by fins arranged to impart a rotational effect on the water entering the propeller which is opposite to the propellers direction and thus generates more thrust. The fins are asymmetrical so as to further even out the inflow of water and counteract the disturbance caused by the hull.

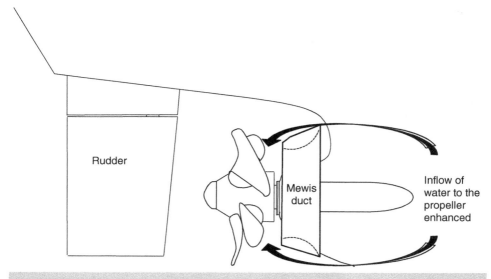

▲ **Figure 8.14** *Mewis duct*

Tail Flaps and Rotating Cylinders

It was stated earlier in this chapter that the rudder angle is usually limited to about 35° on each side of the centreline. It is found that if this angle is exceeded, the diameter of the turning circle is increased, largely due to the separation of the flow of water behind the rudder. In vessels where high manoeuvrability is essential this limit is a disadvantage.

One method of improving the rudder performance is to fit a tail flap which moves automatically to a larger angle as the rudder is turned, in a similar manner to a fin stabiliser (see Chapter 11). Experiments have shown excellent results although the cost of manufacture and maintenance has proved prohibitive until recently. The 'Becker flap' is now finding a market in this time when efficiency gains result in significant fuel savings.

An alternative method which has been tested is to fit a rotating cylinder at the fore end of the rudder. This cylinder controls the boundary layer of water and reduces the separation of the water behind the rudder, producing a positive thrust at larger rudder angles with consequent large reductions in diameter of turning circle and increased rate of turn. Typically, a large tanker travelling at 15 knots would have a cylinder of

1 m diameter rotating at about 350 rev/min and requiring about 400 kW. Although positive thrust is achieved at a rudder angle of 90°, practical considerations would limit the rudder angle to about 70° and even this would require a redesign of steering gears.

Podded Drive

As mentioned earlier in this volume and in others, the podded drive is a significant development for the industry (Figure 8.15). As of 2021 the two major manufacturers in the world for the highest powered pods are Kongsberg and ABB.

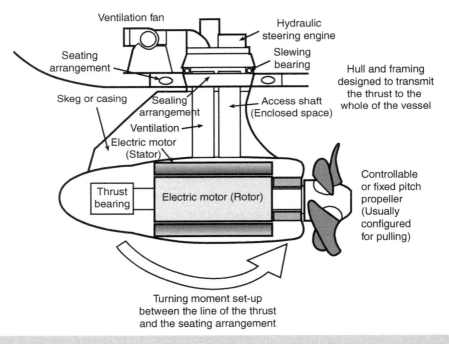

▲ Figure 8.15 *Podded drive*

They vary slightly in design but the basic concept is the same. Both are variable speed electric motors driving a fixed pitch propeller. The largest of the pods are pulling units of around 20 MW power. Large cruise ships might have up to four of these units. They could be arranged so that some are fixed and others provide the steering capability. Siemens-energy also produce large podded drive systems for ocean-going ships. These are interesting as they feature a twin-propeller system, which maximizes the propeller surface area for the lowest weight possible.

The significance of the design is not only in being able to optimise hull efficiency but also in the arrangement of the electrical generators. Free from the constraints of the shaft line requirements the designer can maximise the use of the space available and also optimise the 'cargo' carrying capacity which in the case of the cruise liners will be passenger space.

Where the podded drive is the sole means of propulsion then, generally, it will be good practice to fit at least two units. Any alternative will need to show to the classification society how a sufficient level of redundancy is to be incorporated into the design.

Designers will need to pay particular attention to calculating the load paths setup by the thrusters operating in all their possible directions. A comprehensive 'Failure Modes and Effects Analysis' (FMEA) will need to be assessed and declared to the classification society. In addition, the technical analysis will need to identify any critical components, where failure would result in total loss of propulsion and/or steering.

Looking at Figure 8.15, students will realise that the thrust from the propeller sets up a turning moment that will need to be resisted by the structure of the hull where it joins with the strength part of the pod, which is usually the casing.

The support structure, within the hull of the vessel, will be constructed of a matrix of longitudinal girders, transverse supports and radial girders. These will all be designed to reduce the maximum local stresses and deflection of the hull at the maximum transmission of power from the propulsor.

Cruise ships also require significant power for passenger lighting, galley and air conditioning. Therefore, the 'power station' concept is ideal for matching to the podded drive. See *Reeds Vol 8* for more information about this subject.

9

CONSTRUCTION DETAILS, SPECIFIC TO DIFFERENT TYPES OF VESSELS

Introduction

The International Association of Classification Societies (IACS) has re-examined their rules for the construction of oil tankers and bulk carriers (BCs). Their group of technical experts found considerable overlap and therefore the Common Structural Rules for Bulk Carriers and Oil Tankers has, from 1 July 2015, replaced the existing Common Structural Rules for Double Hull Oil Tankers and the Common Structural Rules for Bulk Carriers.

Oil Tankers

There has been a tremendous growth in the size of tankers since the end of the 1940s. The deadweight of these vessels has increased from about 13 000 tonne in 1946 to the present very large crude carriers (VLCCs) of 150 000–250 000 tonne and the ultra large crude carriers (ULCCs) of over 300 000 tonne, with the largest built to date (2016) being just over 560 000 tonne deadweight. The most significant development in recent years has been the introduction of the 'double hulled' tanker design which was introduced in response to environmental damage that came about due to oil escaping from tankers that had run aground.

The design of the basic oil tanker structure has also progressed quite significantly, once the inherent faults in welded vessels had been overcome by using a continuous, welded structure with well-rounded corners. At the same time improved quality steel, known as *extra-notch tough steel*, has been introduced which is less susceptible to the formation of cracks and will also resist the spread of cracks.

Following on from the basic 'single hulled' tankers there were variations such as the double bottom design, the double sided version and now the full double hulled design. However, the first significant development was in the way that the cargo tanks were cleaned ready for the next load.

Originally it was common practice to wash the tanks in crude oil carriers with water, pump the oily residue into the sea and fill the same tanks with water 'ballast'. Upon arrival at the loading port, and sometimes before arrival, the ballast was pumped out into the sea. The water washing was done using powerful jets; however following serious explosions it was found that the water jet was generating static electricity which in turn generated sparks initiating an explosion as the tank still contained a volatile atmosphere.

Two developments happened subsequently; the first was the introduction of inert gas in the space above the cargo and the second was the development of 'crude oil washing' (COW).

Tankers were then required to have segregated ballast systems so that the oil cargo was always separated from the ballast water. The latest vessels are required to have a full double hull as well as a ballast water treatment system with the latter designed to ensure that the ship does not transfer marine life which is common to one part of the world to another place where it is not a native species.

The basic structural arrangements and details vary considerably from shipyard to shipyard, but there are two basic methods of framing that have been used. These are:

1. Longitudinal framing, in which the deck, bottom shell, side shell and longitudinal bulkheads are stiffened longitudinally (Figure 9.1).
2. Combined framing, in which the deck and bottom shell are framed longitudinally, with transverse side frames and vertical stiffeners on the longitudinal bulkheads (Figure 9.2).

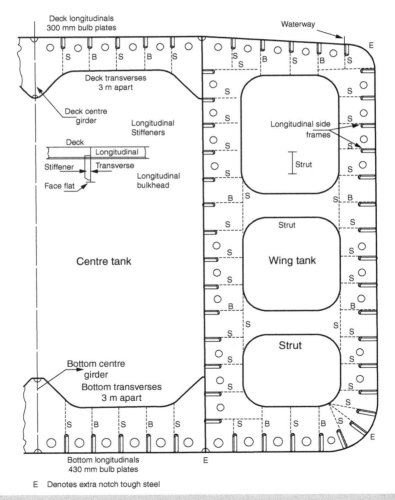

▲ **Figure 9.1** *Oil tanker – longitudinal framing*

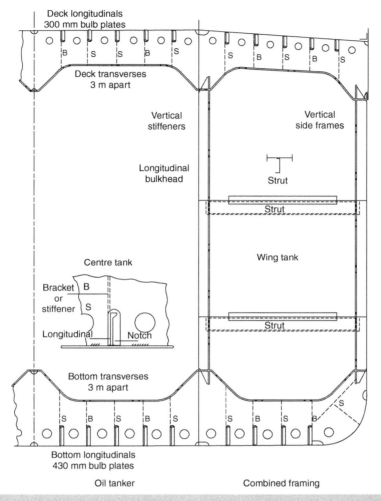

▲ **Figure 9.2** *Oil tanker – combined framing*

The longitudinal framing method has now become the dominant method.

Longitudinal framing

The longitudinal framing system provides ample longitudinal strength, but the horizontal side frames and longitudinal bulkhead stiffeners are likely to retain liquid and hence will be susceptible to increased corrosion. The longitudinals are supported by deep transverse webs which form rings round the ship at intervals of 3–6 m and to which they are attached by flat bars or brackets. At the transverse bulkheads, the structure must be carefully designed to give maximum continuity of strength. Figure 9.3 shows typical attachments of bottom and side longitudinals at the transverse bulkhead.

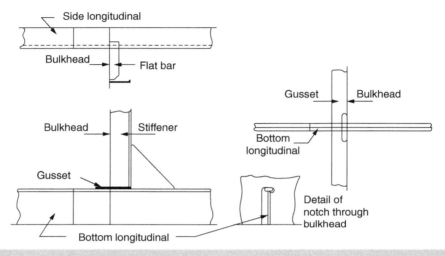

▲ **Figure 9.3** *End connections of longitudinal*

The longitudinals are usually bulb plates although many of the larger vessels employ large flat plates. The bottom longitudinals are much heavier than the deck longitudinals, while the side longitudinals increase with the depth of the tank. The transverse webs fitted to the side shell and longitudinal bulkhead are strengthened by face flats and supported by two or three horizontal struts arranged in such a way that the unsupported span of transverse at the top is greater than that at the bottom, the latter being subject to a greater head of liquid. It is intended by this form of design that the bending moments on the separate spans should be equal. It is essential that the face flat on the web is carried round the strut to form a continuous ring of material.

Combined framing

This system has proved successful in many ships, having the advantage of providing sufficient longitudinal strength with good tank drainage due to the vertical side frames and stiffeners. The latter are supported by horizontal stringers which are continuous between the transverse bulkheads and tied at intervals by struts. The lower stringers are heavier than the upper stringers. Where the length of the tanker exceeds 200 m Lloyd's require longitudinal framing to be used.

General structure

A deep centreline girder must be fitted at the keel and deck, connected to a vertical web on the transverse bulkhead. This web is only required on one side of the bulkhead. The top and bottom girders act as supports for the transverse structures and hence reduce their span. Large face flats are fitted to their free edges and continued round the bulkhead web to form almost a complete vertical ring (Figure 9.4). The bottom centre girder is required to support the ship while in dock and it is found necessary for this reason to reduce the unsupported panels of keel plate by fitting docking brackets between the transverse structures, extending from the centre girder to the adjacent longitudinal (see Chapter 2). In many of the larger ships the centre girders have been replaced by a full-height perforated bulkhead.

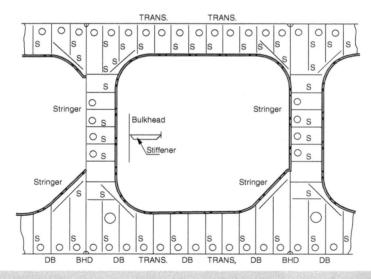

▲ **Figure 9.4** *Centreline web*

The transverse bulkheads are usually stiffened vertically, the stiffeners being bracketed at their ends and supported by horizontal stringers (Figure 9.5). Corrugated bulkheads are often fitted, the corrugations being vertical, and have the advantages of improving drainage, allowing easy cleaning and reducing weight. The longitudinal bulkheads may also be corrugated but in this case the corrugations must be horizontal, otherwise

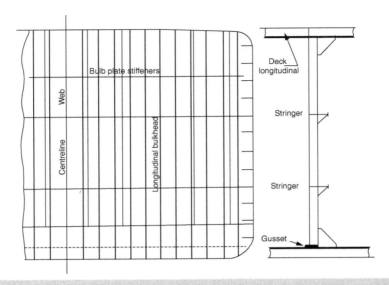

▲ **Figure 9.5** *Oil-tight bulkhead*

the longitudinal strength will be impaired as the bulkhead would tend to fold like a concertina as the ship hogs and sags.

The thickness of the deck plating depends on the maximum bending moment to which the ship is liable to be subject, and is given in the form of a cross-sectional area of material with respect to openings. These openings consist of oil-tight hatches and tank measuring equipment and/or openings through which deep-well pumps could be inserted (Figure 9.6).

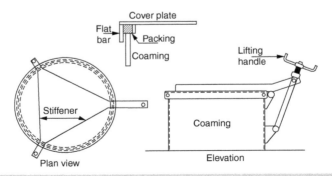

▲ **Figure 9.6** *Oil-tight hatch*

Throughout the ship the greatest care is taken to ensure continuity of the structure which maintains the girder strength of the vessel. At the ends, the longitudinals reduce in number gradually, however, in the engine room and at the fore end, the ship is given additional strength transversely to supplement the longitudinal strength. The ends of the longitudinal bulkheads could be continued in the form of brackets, while in some designs the bulkheads are carried through the whole length of the machinery space, forming tanks, stores and workshop spaces at the sides.

Cargo pumping and piping arrangements

An overview of piping and pumping arrangements are covered in this volume as it is necessary to include the different methods of loading and discharging oil cargoes. Traditionally a cargo pump room is arranged at the after end of the cargo tank range, containing three or four large capacity centrifugal pumps together with between two and four smaller capacity stripping pumps. These pumps are used to clear the tanks of oil when the main cargo pumps lose suction. In addition, two ballast pumps may be provided, especially where segregated ballast systems are used.

Several systems of piping are in use, depending largely on whether it is intended to carry single grade or multi-grade cargo. A simple piping arrangement may be used for single grade cargo as shown in Figure 9.7. This is known as a ring main system and consists of a continuous pipe from the pump room to the forward cargo tank and back into the pump room, with cross-over pipes in each centre tank extending into the wing tanks. This system may be used in conjunction with two or three pumps, the centre pump assisting one or both of the outside pumps. It is possible to carry two different grades of cargo with this system and to discharge them both at the same time. Valves with extended spindles up to the deck are fitted in each tank, allowing the tank to be discharged or by-passed as required. The oil is discharged through pipes fitted on the upper deck amidships and led aft along the deck to the pump room, while a single stern discharge might be arranged.

A more complicated piping arrangement is required when several grades of cargo are carried and one such arrangement is shown in Figure 9.8. In this case four main cargo pumps may be used to discharge the separate cargoes which are drawn from the tanks through individual lines. Two smaller capacity stripping pumps are fitted with connections to the main cargo tank lines.

A more modern arrangement has been developed in which the oil is allowed to flow through hydraulically operated sluice valves into a common suction strum from where it is discharged by the main cargo pumps. This system may be used for single grade

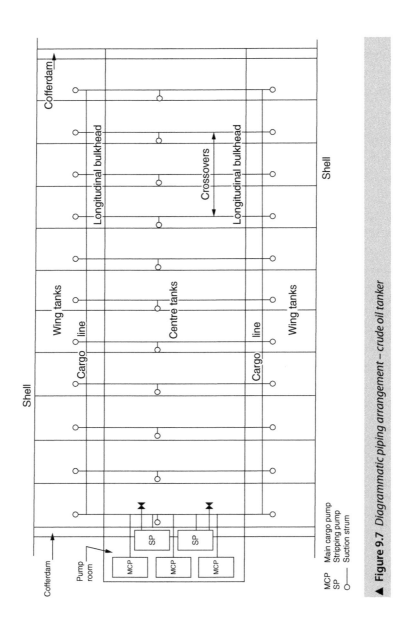

MCP   Main cargo pump
SP   Stripping pump
○—— Suction strum

▲ **Figure 9.7** *Diagrammatic piping arrangement – crude oil tanker*

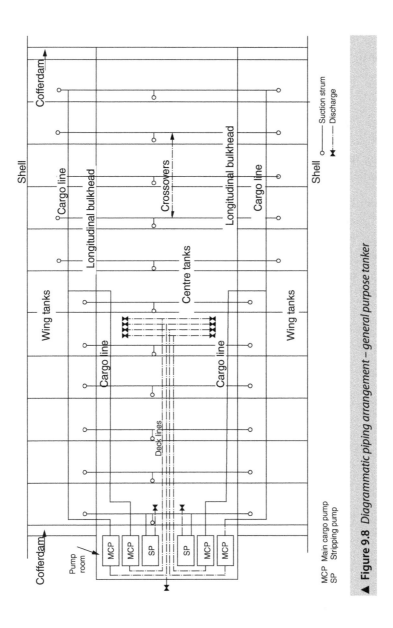

▲ **Figure 9.8** *Diagrammatic piping arrangement – general purpose tanker*

cargo and is intended to dispense with the cargo pipelines in the tanks. In case of failure of the valves, however, a heavy duty stripping pump and line is fitted.

Many oils must be heated before discharging and therefore heating coils are fitted to all tanks. The coils are of cast iron with large heating surfaces which are heated by means of steam. It is essential that these coils be close to the bottom of the tank, and it is therefore necessary to cut holes in the bottom transverses to allow the coils to be fitted.

While many different types of pump are available, several ships make use of vertical drive pumps in which the driving spindles extend through an oil-tight flat into the machinery space, the motors being fitted on the flat. This ensures regular maintenance by the engine room staff without entering the pump room.

When cargo is being loaded or discharged using the high capacity pumps, the structure is subject to an increase or reduction in pressure. Similar variations in pressure occur due to the thermal expansion or contraction of the cargo. To avoid structural damage, some form of pressure/vacuum release must be fitted to accommodate these variations. In addition, provision must be made to ventilate the gas-filled space, ensuring that the gas is ejected well above the deck level. For many years the latter was achieved by running vapour lines from the hatches to a common vent pipe running up the mast and fitted with a flame arrester at the top. More recently individual high velocity vents on stand pipes have been fitted to the hatches. By restricting the orifice, a high gas

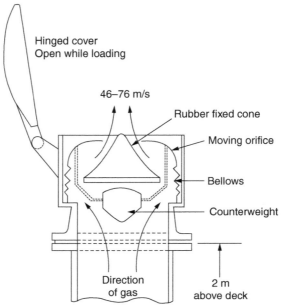

▲ **Figure 9.9** *Cargo gas venting system*

velocity is produced when loading. A pressure/vacuum valve is incorporated in the system. Figure 9.9 shows a typical system in the open position.

In order to reduce the possibility of explosion in the cargo tanks the oxygen content is reduced by means of an inert gas system. It is essential that the tank space above the liquid is inerted at all times, which means, when the tanks are full, empty or partially filled and particularly during the tank cleaning operation.

Deepwell pumping systems

With this, the most modern of systems, each tank has its own 'deepwell' pump which replaces the large centralised cargo pumps. The electric or hydraulic motor is arranged at the top of each pump at the deck level. Each motor sits above the pump head which is working at the bottom of the tank and is connected via a long drive shaft.

The electrically driven pumps may also be fitted with variable speed electric motors, as standard or retrofitted, making them particularly efficient and suited for this purpose. The hydraulic motors will also have the ability to vary the speed of the pump and, thus, vary the speed/volume of pump delivery.

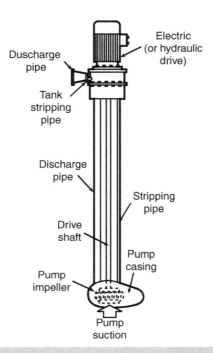

▲ **Figure 9.10** *Deepwell pump*

Another advantage of this type of pump is that the impeller is operating in the medium (for example, the crude oil) to be pumped. This means that it will not have problems with losing suction and will, therefore, make pumping the final litres of oil (a process known as stripping the tanks) easier.

Crude oil washing

As the cargo is being discharged the tanks can be cleaned as the level falls. The sludge, which accumulates on horizontal surfaces and in the bottom of the tanks, can then be discharged along with the cargo or, alternatively, discharged into a holding tank ready for disposal to a shore facility. No oil is allowed to be discharged into the sea/river/ dock. An older and now banned practice was for the tanks to be water washed using high pressure nozzles from either portable or fixed machines. Even when using hot water, however, not all the sludge was removed and it was necessary at times to dig out the heavy deposits. In addition, it is suspected that water washing contributed to the explosions on three VLCCs during the ballast passage while washing cargo tanks.

Crude oil is itself a solvent and is now used as a washing agent as an alternative to water in crude oil carriers. The crude oil washing (COW) system comprises a permanent arrangement of steel piping, independent of the fire mains or any other piping system, connected to high capacity washing machines having nozzles which rotate through 120° and project the crude oil at about 9–10 bar. The supply to the machines is provided either by the main cargo pumps or by pumps provided for that purpose.

The machines are so located that all horizontal and vertical areas within the tank are washed either by direct impingement or by deflection or splashing of the jet. The number and disposition of the machines will depend upon the structural arrangements within the tank, but less than 10% of the total surface of the structure may be shielded from direct impingement.

It is essential that an efficient inert gas system is used in conjunction with the COW system. It is also a requirement that only trained personnel carry out the work and a comprehensive operations and equipment manual is required on each ship.

There were some advantages over the water washing system, however modern environmental considerations dictate that this is now the only practice as all tankers and gas carriers must not discharge any oil into the sea water.

Tanker vetting – Chemical Distribution Institute (CDI) and their Ship Inspection Reporting System (SIRE) – is an inspection scheme used to ensure that tankers that are being chartered conform to all the necessary rules and regulations that are in-force at the time. This means that 'charters' can be sure that they are hiring a robust vessel.

Bulk Carriers

Originally the majority of bulk cargoes were carried by general cargo ships which were not designed specifically for such cargoes. This was inefficient and led to the development of the specialised BC. The rate of loading in bulk at some terminals is tremendous, 4000–5000 tonne/hour being regarded as normal.

It used to be that the design of a BC depended a great deal on the type of cargo that is to be carried and which tanks, if any, were to remain empty. However, high density cargoes such as iron ore will require a high strength vessel since the ore is heavy for a given volume and other cargoes will be less dense. This led the Maritime Safety Committee (MSC85) of the International Maritime Organization (IMO) to consider the definition of a BC in November/December 2008.

The BC could be constructed as single or double skin ships with double bottoms and have hopper side tanks and topside tanks. From the early part of 2015 BCs had to be built to the IACS 'Common Constructional Rules for Tankers and Bulk Carriers'. These rules apply to BCs capable of unlimited voyages and having their engine rooms situated aft of the cargo tanks.

The general assumptions for the IACS rules are that the ships are designed, constructed and operated under the rules set out by flag states and a recognised classification society. The safety, watertight subdivisions, stability criteria and fire protection all conform to the recognised standards. The design life and strength of the vessel are adequate for the intended service of the ship.

However, as stated, the design of a bulk carrier must be of high strength due to the high-density cargoes it is required to carry. For example, ore is heavy for a given volume and therefore a deep double bottom is fitted in the vessel that carries it.

A deep double bottom is fitted, together with longitudinal bulkheads which restrict the ore and maintain a high centre of gravity consistent with comfortable rolling. A ship designed to carry bauxite, however, requires twice the volume of space for cargo and will therefore have a normal height of double bottom although longitudinal bulkheads may be used to restrict the ore space.

A bulk 'tramp' if one can coin a term, that is, one which may be required to carry any type of bulk cargo, must have restricted volume for an iron ore cargo and at the same time must have sufficient cargo capacity to carry its full deadweight of light grain which requires 3 times the volume of the ore. One method of overcoming this difficulty is to design the ship to load ore in alternate holds.

It may readily be seen, therefore, that the design of BCs will vary considerably. However, the practice of operating a bulk ore carrier (OBO) vessel (Figure 9.11) has died out. The figure shows a cross-section of an OBO which may carry an alternative cargo of oil in the wings and double bottom.

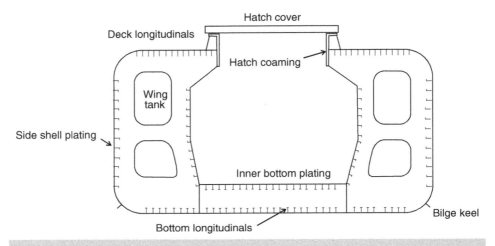

▲ **Figure 9.11** *Ore carrier – mid-section*

The structure is similar to that required for oil tankers, having longitudinal framing at the deck, bottom and side shell, longitudinal bulkhead and tank top. These longitudinals are supported by transverse webs 2.5 m apart. The supporting members are, as far as possible, fitted in the tanks rather than the ore space, to facilitate discharge of cargo using grabs. For the same reason it is common practice to increase the thickness of the tank top beyond that required by the classification societies.

The structural arrangement of a BC is shown in Figure 9.12. The arrangement of the stiffening members is once again similar to the oil tanker, although the layout of the cargo space is entirely different from that shown in Figure 9.11. In this design the double bottom, lower hopper and upper hopper spaces are available for water ballast, the upper tanks raising the centre of gravity of the ship and hence reducing the stiffness of roll. A large cargo capacity is provided by the main cargo holds and, if necessary, the upper hopper tanks.

Liquefied Gas Carriers

It is now becoming much more popular to carry natural gas and petroleum gas in liquefied form rather than as a vapour, the volume of the liquid being one-three

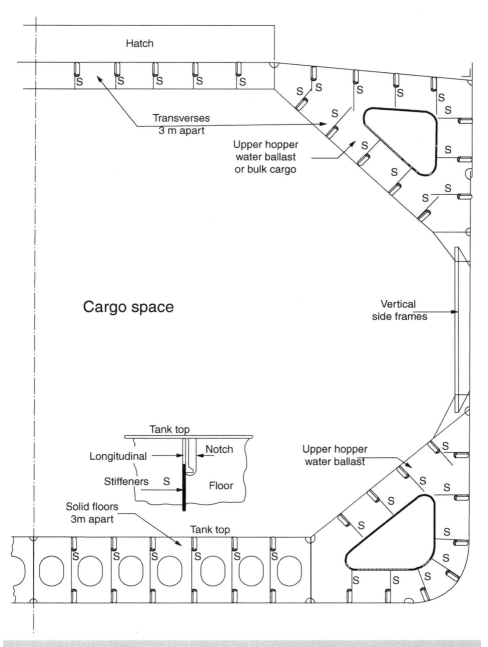

▲ **Figure 9.12** *Bulk carrier*

hundredth to one-six hundredth of the volume of the vapour. Several gases are transported in this way, such as methane, propane, butane and anhydrous ammonia. The gases are divided into *liquefied natural gas* (LNG), which consists mainly of methane, and *liquefied petroleum gas* (LPG), mainly propane, propylene, butane and butylene. The latter are derived from the refining of the LNG or as a by-product of the distillation of crude oil at the refineries.

Methane has a boiling point at atmospheric pressure of −162°C, while its critical temperature is −82°C at a pressure of 47 bar, that is, the gas cannot exist as a liquid at a temperature higher than −82°C, no matter what the pressure is. Thus the containment system for LNG must be suitable for conditions between these limits. It is usually more economical to design for the lower temperature at atmospheric pressure.

LPG requirements vary between maximum required pressures of about 18 bar to atmospheric pressure, and minimum temperatures of about −45°C (ethylene −104°C) to −5°C. Many different types of tank systems have been introduced and may be considered under three main headings.

Fully pressurised

The tanks are in the form of pressure vessels, usually cylindrical (Figure 9.13) designed for a maximum pressure of about 18 bar. No re-liquefaction plant is required and no insulation is fitted. Relief valves protect the tank against excess pressure. A compressor is fitted to pressurise the cargo. This tank system has not proved popular due to the considerable loss in hold capacity, the weight of the system and the subsequent cost. Fully pressurised tanks have usually been fitted to small ships having a capacity of less than 2000 m³. The tanks must be strongly built and are termed *self-supporting*.

Semi-pressurised/partly refrigerated

In order to reduce the cost of the tank system, the tank boundaries may be insulated and a re-liquefaction plant fitted. The maximum pressure is about 8 bar and the minimum temperature about −5°C. This system has been used on vessels up to about 5000 m³ capacity. The tank system is similar in form to the fully pressurised tanks and has the same loss in hold capacity.

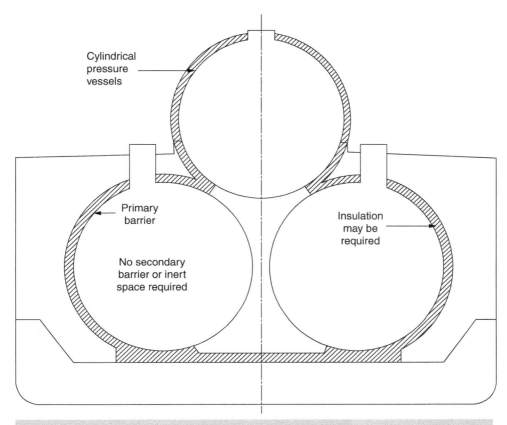

Cylindrical
pressure
vessels

Primary
barrier

No secondary
barrier or inert
space required

Insulation
may be
required

▲ **Figure 9.13** *Cylindrical tank system*

Semi-pressurised/fully refrigerated

This tank system is designed to accept pressures of about 8 bar, but is built of material which will accept temperatures down to about −45°C. The structure must be well-insulated and a re-liquefaction plant is necessary. The tanks may be cylindrical or spherical (Figure 9.14) and are self-supporting. Ships of this type may carry cargoes under a range of conditions, from high pressure at ambient temperature to low temperature at atmospheric pressure.

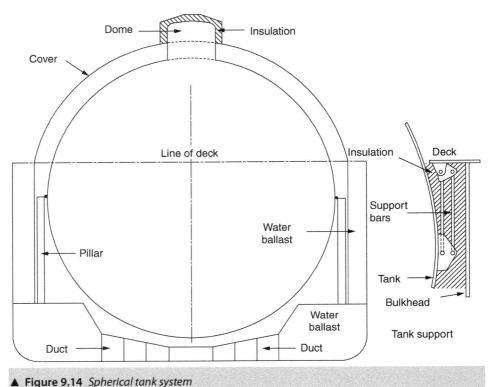

▲ **Figure 9.14** *Spherical tank system*

Fully refrigerated

Cargo is carried at atmospheric pressure and at a temperature at or below the boiling point. The system is particularly suitable for the carriage of LNG but is also used extensively for LPG and ammonia. Vessels designed for LNG do not usually carry a re-liquefaction plant but LPG and ammonia vessels may gain from its use. The tank structure may be of prismatic form (Figure 9.15) or of membrane construction.

Prismatic tanks are self-supporting, being tied to the main hull structure by a system of chocks and keys (Figure 9.16). They make excellent use of the available space.

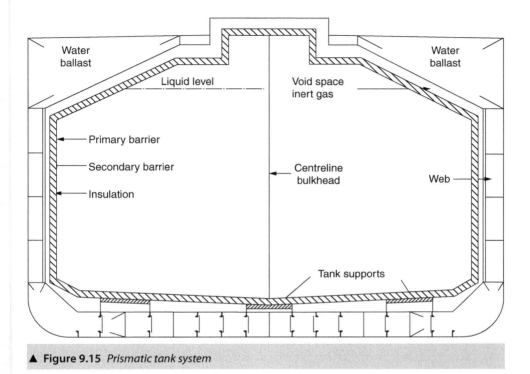

▲ **Figure 9.15** *Prismatic tank system*

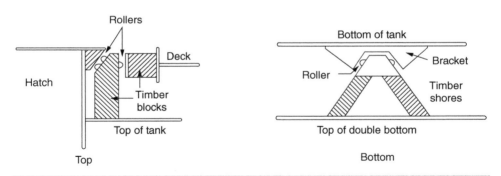

▲ **Figure 9.16** *Tank chocks*

Membrane tanks are of rectangular form and rely on the main hull structure for their strength. A very thin lining (0.5–1.2 mm) contains the liquid. This lining must be constructed of low expansion material or must be of corrugated form to allow for changes in temperature. The lining is supported by insulation which must therefore be load-bearing (Figure 9.17).

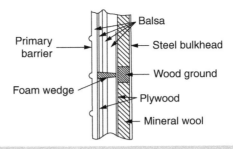

▲ **Figure 9.17** *Membrane system*

For both membrane and prismatic tanks having a minimum temperature less than −50°C either nickel steel or aluminium must be used. In both cases secondary barriers are required if the minimum temperature is less than −10°C. Thus, in the event of leakage from the primary container, the liquid or vapour is contained for a period of up to 15 days. If the minimum temperature is higher than −50°C, the ship's hull may be used as a secondary barrier if constructed of Arctic D steel or equivalent. Independent secondary barriers may be of nickel steel, aluminium or plywood as long as they can perform their function.

Several types of insulation are acceptable, such as balsa wood, polyurethane, pearlite, glass wool and foam glass (Figure 9.17). Indeed, the primary barrier itself may be constructed of polyurethane which will both contain and insulate the cargo. Usually, however, the primary barrier is of low-temperature steel or of aluminium, neither of which become brittle at low temperatures.

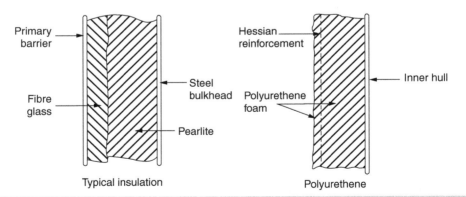

▲ **Figure 9.18** *Tank insulation*

Care must be taken throughout the design to prevent the low-temperature liquid or vapour coming into contact with steel structure which may become brittle and hence fracture.

Safety and environmental control

Due to variations in external conditions and the movement of the ship the liquid cargo will boil and release gas. This gas must be removed from the tank to avoid increased pressure. If the gas is lighter than air it may be vented to the atmosphere, vertically from the ship and away from the accommodation. Many authorities are concerned about the increased pollution and alternatives are encouraged (see 'Boil-off' below).

The cargo piping system is designed for easy gas freeing and purging, with gas sampling points for each tank. The spaces between the barriers and between the secondary barrier and the ship side are either constantly inserted or sufficient inert gas is made available if required.

Each tank is fitted with means of indicating the level of the liquid, the pressure within the tank and the temperature of the cargo. A high liquid level alarm is fitted, giving both audible and visible warning, and automatically cuts off the flow.

High and low pressure alarms are fitted within the tanks and in the inter-barrier space if that space is not open to the atmosphere. A temperature measuring device is fitted near the top and at the bottom of each tank with an indicator showing the lowest temperature for which the tank is designed. Temperature readings are recorded at regular intervals, while an alarm will be given if the minimum temperature is approached.

Gas detection equipment is fitted in inter-barrier spaces, void spaces, cargo pump rooms and control rooms. The type of equipment depends upon the type of cargo and the space in question. Measurements of flammable vapour, toxic vapour, vapour and oxygen content are taken, audible and visible alarms being actuated if dangerous levels are recorded. Measurements for toxic gas are recorded every 4 h except when personnel are in the compartment, when 30 min samples are taken and analysed.

In the event of a fire, it is essential that the fire pumps are capable of supplying two jets or sprays which can reach all parts of the deck over the cargo tanks. The main fire pumps or a special spray pump may be used for protecting the cargo area. The sprays may also be used to reduce the deck temperature during the voyage and hence reduce the heat gain by the cargo. If the cargo is flammable, then a fixed dry chemical fire extinguishing system is fitted.

Boil-off

In the early stage of LNG design, excess methane gas was vented astern of the ship and burned as it exited to the atmosphere. As the LNG vessel became more popular and sophisticated and as re-liquefaction was not economical, the excess gas was used as fuel for the main engine.

This is still common practice and in motor vessels normal injection equipment is used, together with a hydraulically operated gas injection valve in the cylinder head, blowing in the 'boil-off' gas at about 3 bar into the incoming scavenge air. The gas line from the tanks is fitted with a relief valve. Under normal running conditions the gas is used as the fuel. Should the gas pressure fall, however, liquid fuel is injected, a 10% drop in pressure resulting in automatic oil supply, while if the gas pressure falls by 15% the gas flow is stopped. Oil fuel must be used when starting or manoeuvring.

In steam ships the firing equipment is capable of burning oil fuel and methane gas simultaneously. The oil flame must be present at all times and an alarm system is fitted to give warning of pump failure causing loss of oil. The nozzles are purged with inert gas or steam before and after use. As with motor ships, oil fuel is used when starting or manoeuvring.

Currently any excess gas in LNG ships is used in the main engine or re-liquefied by passing it through a cooling system and returning it to the tanks in liquid form. Methane is a highly potent 'greenhouse' warming gas and it is highly undesirable, and against all international regulations, to release it into the atmosphere unless it is an emergency.

Operating procedures

Drying

Water in any part of the cargo handling system will impair its operation, by freezing, reducing the purity of the cargo or in some cases changing its nature. The system must be cleared of water or water vapour by purging with a dry gas or by the use of a drying agent.

Inerting

If the oxygen content in a tank is too high a flammable mixture may be produced or the oxygen may be absorbed by the cargo producing a chemical change. It is essential, therefore, to reduce the oxygen content by the introduction of inert gas to a maximum content of 5% for hydrocarbons, 12% for ammonia and 0.5% for ethylene. The most suitable and common inert gases are nitrogen and carbon dioxide and in some cases helium, argon and neon are used but these are expensive. Inert gas generated by the ship-board plant usually consists of about 84% nitrogen and 12%–15% carbon dioxide with an oxygen content of about 0.2%. The inert gas, in turn, must be purged from the system by the cargo gas vapour. An inert gas barrier must also be used when discharging cargo and allowing air into the tanks.

Pre-cooling

Classification societies require that the maximum temperature difference between the cargo and the tank should not exceed 28°C. Before loading, the tank may be at ambient temperature. It is then cooled by spraying liquefied gas into the tank. The gas then vaporises and cools down the tank. The vapour produced in this way is either vented or re-liquefied. The cooling rate must be controlled to prevent undue thermal stresses and excess vapour and a rate of between 3°C and 6°C per hour is usual.

Loading and discharging

The cargo pipeline must first be cooled before loading commences. The rate of loading depends largely on the rate at which the cargo vapour can be vented or re-liquefied. Thus a ship designed for a particular run should have a re-liquefaction plant compatible with the loading facilities.

When the cargo is discharged it must be in a condition suitable for the shore-based tanks. Thus, if the shore tanks are at atmospheric pressure, the cargo in the ship's tanks should be brought to about the same pressure before discharging commences.

Sufficient net positive suction head must be provided for the pumps to work the cargo without cavitation. In pressure vessels the ship's compressor may pressurise the vapour from an adjoining tank to maintain a positive thrust at the impeller. In refrigerated ships the impeller is fitted at the bottom of the tank, the liquid head producing the pressure required. In the event of pump failure, the cargo may be removed by the injection of inert gas. Booster pumps are usually fitted to overcome any individual pump problem and ensure a continuous rate of discharge.

Container Ships

Container ships are designed to carry large numbers of standard containers. When first introduced the ships were designed for high speeds between specific terminal ports and they required a fast turn round at those ports.

The containers are of international standard either 20 ft or 40 ft in length, 8 ft in width and 8 ft 6 in in height. There is a less popular 'Hi-Cube' container which is of a standard size but 9 ft 6 in high. These can be straight containers or they can be refrigerated.

Containers are strong enough to be stacked six high. There are two basic types of refrigerated containers available, one which carries its own refrigeration plant, either fixed or clipped on, and one which relies on air from brine coolers in the ship which is ducted to the container. The refrigerated container with the 'clip-on' unit is the most popular as it can easily be transported across 'intermodal' transport systems. The cross-section of a container ship, showing its construction details, can be seen in Figure 9.19.

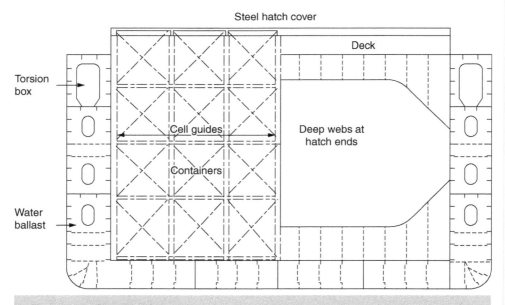

▲ **Figure 9.19** *Container ship*

The containers are loaded into the ship vertically, fitting into the top of the container underneath or sliding into cell guides which are chamfered out at the top to provide a lead-in. Pads are fitted to the tank top at the bottom of the guides in line with the corner fittings. The available hold space is dictated by the size of the hatches. It is essential, therefore, to have long, wide hatches to take a maximum number of containers. The spaces at the sides of the hatch are used for access and water ballast. The hatch coamings and covers are designed to carry tiers of containers as deck cargo. Since the vessels mostly work between well-equipped ports, they do not usually carry their own cargo handling equipment.

Due to the wide hatches the deck plating must be thick, and higher tensile steel is often used. The deck, side shell and longitudinal bulkheads are longitudinally framed in addition to the double bottom. The hatch coamings may be continuous and therefore improve the longitudinal strength. Problems may arise in these vessels due to the lack of torsional strength caused by the large hatches. The girder strength is boosted by fitting additional scantlings which act as *torsion boxes* on each side of the ship. These boxes are formed by the upper deck, top part of the longitudinal bulkhead, sheerstrake and upper platform, all of which are of thick material. The boxes are supported inside by transverses and wash bulkheads in addition to the longitudinal framing.

These transverse boxes are only effective if they are efficiently secured at their ends. At the after end they extend into the engine room and are tied to deep transverse webs. Similarly at the fore end, they are carried as far forward as the form of the ship will allow and are welded to transverse webs. The longitudinal bulkheads below the box may have to be stepped in-board to suit the shape of the ship, the main longitudinal bulkhead being scarphed into the stepped section.

At the ends of the hatches deep box webs are fitted to increase the transverse and torsional strength of the ship. These webs are fitted at tank top and deck levels. Care is taken in the structural design at the hatch corners to avoid excessive stresses.

The double bottom structure beneath the cell guides is subject to impact loading as the containers are put on board. Side girders are usually fitted under the container seats with additional transverse local stiffening to distribute the load. Unlike normal cargo ships in which the cargo is distributed over the tank top, the inner bottom of a container ship is subject to point loading. The double bottom must be deep enough to support the upthrust from the water when the ship is deeply loaded, without distortion between the containers.

The 'liner' trade is now dominated by container ships and the size is generally measured in their ability to carry containers. The 20 ft container is used as a measuring scale in the form of the twenty foot equivalent unit (TEU) and ships of 20,000 TEU are now being planned.

10

THE LOAD LINE REGULATIONS

The general introduction to this book explains how International Shipping has become increasingly guided by the decisions made within the International Maritime Organization (IMO), which is the the maritime department of the United Nations (UN).

The International Convention on Load Lines was adopted by the IMO in 1968. Previously, the rules for the construction and loading of ships was left to the individual flag administrations. The updated version that entered into force in February 2000 harmonised the convention with Safety Of Life At Sea (SOLAS) and the International Convention for the Prevention of Pollution from Ships (MARPOL). This chapter gives a brief introduction to the important features of the convention that help to keep ships safe and fit for purpose.

Freeboard

Originally associated with 'the Plimsoll line', *freeboard* is the distance from the waterline to the top of the deck plating at the side of the freeboard deck amidships. The *freeboard deck* is the uppermost continuous deck (also known as the 'bulkhead deck', see page 32) that has the necessary equipment to close all openings to the outside weather.

The *minimum freeboard* is based on providing the vessel with a volume of reserve buoyancy which cannot be loaded with cargo and therefore may be regarded as making the ship safe and ensuring that the ship proceeds to sea in a stable condition.

However, the exact level of 'reserve buoyancy' required depends upon several factors:

- conditions of service of the ship
- type of vessel
- stability of the vessel in still water
- degree of subdivision after suffering 'prescribed damage'
- safety of the ship's staff when out on deck
- ability of the vessel to protect the weather deck from taking on water
- fixtures and fittings used to allow any 'shipped' water to be removed.

In deep sea ships, for example, sufficient reserve buoyancy must be provided to enable the vessel to rise up again when shipping the heavy seas that could be encountered in the oceans of the world, small vessels on the inland waterways will not encounter such conditions and therefore are allowed to sail with a different level of 'reserve buoyancy'.

Vessels conforming to the Load Line Rules are assigned a freeboard according to a table of values, and it is therefore termed the *tabular freeboard*. The initial table used depends upon the type of ship and its length and is based on a standard vessel having a block coefficient of 0.68, length ÷ depth of 15 and a standard sheer curve.

Corrections are then made to this value for any variation from the standard, together with deductions for the reserve buoyancy provided by weather tight superstructures on the freeboard deck. One further point to consider is the likelihood of water coming onto the fore deck. This is largely a function of the distance of the fore end of the deck from the waterline and for this reason a *minimum bow height* is stipulated.

The bow height required depends upon the length of the ship and the block coefficient and may be measured to the forecastle deck if the forecastle is 7% or more of the ship's length. Should the bow height be less than the minimum then either the freeboard is increased or the deck raised by increasing the sheer or fitting a forecastle. The function of the forecastle has been brought into sharp focus recently as some bulk carriers suffered weather damage to the forward hatch coaming due to the lack of protection offered by the inadequate forecastle.

There are two sets of tables (A and B) and students who are familiar with looking at loaded ships will know that, for example, an oil tanker fully loaded will look much lower in the water than will a car carrier.

Type A ships are designed to carry only liquid cargoes and hence have a high integrity of exposed deck, together with excellent subdivision of the cargo space. The hatches are small openings and are oil/water tight, and heavy seas are unlikely to cause flooding of the cargo space or the accommodation. As a result, these vessels are allowed to load to a comparatively deep draught than the type B ships.

While these ships have a high standard of watertight deck, they have a comparatively small volume of reserve buoyancy and may therefore be less safe if damaged. It is necessary, therefore, in all such vessels over 150 m in length, to investigate the effect of damaging the underwater part of the cargo space and, in longer ships, the engine room. Under such conditions the vessel must remain afloat without excessive heel and have positive stability.

Type B ships cover the remaining types of vessels and are assumed to be fitted with steel hatch covers. In older ships having wood covers the freeboard is increased.

Should the hatch covers in Type B ships be sealed with efficient securing arrangements, then their improved water tight integrity is rewarded by a reduction in freeboard of up to 60% of the difference between the Type A and Type B tabular freeboards.

If, in addition, the vessel satisfies the remaining conditions for a Type A ship (e.g. flooding of cargo spaces and engine room), 100% of the difference is allowed and the vessel may be regarded as a Type A ship.

The tabular freeboards for Types A and B ships are given in the Rules for lengths of ship varying between 24 m and 365 m.

Typical values are as follows:

Length of ship	Type A	Type B	Difference
m	Freeboard in mm		
24	200	200	–
100	1135	1271	136
200	2612	3264	652
300	3262	4630	1368
365	3433	5303	1870

The freeboard, calculated from the tabular freeboard and corrected, is termed the Summer Freeboard and corresponds with a Summer Load Line S. In the tropics the weather is usually favourable and therefore a deduction of $\frac{1}{48}$ the Summer draught may be made to give the Tropical Load Line T. Similarly there is a Winter Load Line W which is a penalty of $\frac{1}{48}$ the Summer draught. In ships 100 m and less in length there is a further penalty of 50 mm if the vessel enters the North Atlantic in the Winter (WNA).

The above freeboards are based on the assumption that the ship floats in sea water of 1025 kg/m³. If the vessel floats in fresh water with the same displacement, then it will lie at a deeper draught. The *Fresh Water Allowance is* $\frac{4}{41}$ mm where displacement is the measurement taken in sea water at the Summer draught and T is the tonne per cm at the same draught. F represents the fresh water line in the Summer zone and TF the equivalent mark in the Tropical zone.

The freeboard markings (Figure 10.1) are cut into the shell plating with the centre of the circle at midships. The letters LR on the circle indicate that the load line has been assigned by Lloyd's Register of Shipping. If the vessel has a radiused gunwale, the deck line is cut at a convenient distance below the correct position and this distance is then deducted from the freeboard stated on the Load Line Certificate.

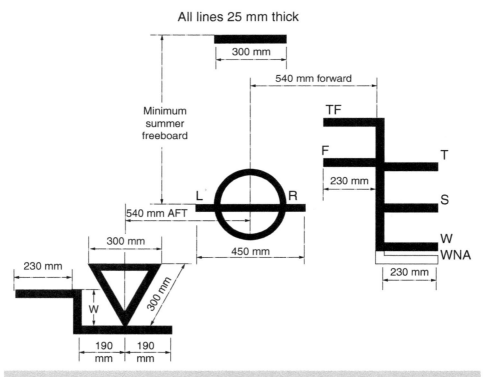

▲ **Figure 10.1** *Freeboard markings and tonnage mark*

Special provision is made in the Rules for vessels carrying timber as a deck cargo. The timber increases the reserve buoyancy and hence the vessels are allowed to float at deeper draughts. An additional set of freeboard markings is cut in aft of midships with the normal letters prefixed by L (lumber).

Conditions of assignment

The Load Line Rules are based on the very reasonable assumption that the ship is built to and maintained at a high level of structural strength and will sail in a safe and seaworthy condition. However, the legal definition of seaworthiness is quite subjective and could be open to interpretation in a court of law.

Until recently the Rules laid down the standard of longitudinal and transverse strength. The classification societies usually found it necessary to increase these standards although in some designs the Rules were considered excessive. It is now felt that the structural strength of the ship is more properly the function of the classification societies who may well be the Assigning Authority.

Standards of stability are given in the Rules for both small and large angles of heel. Details of the information required to be carried on a ship are stated, together with typical calculations. All the information is based on an inclining experiment carried out on the completed ship in the presence of a Flag surveyor.

It is essential that all openings in the weather deck are weathertight. Hatch coamings, hatch covers, ventilator coamings, air pipes and doors must be strong enough to resist the pounding from the sea and standards of strength are given in the Flag/Classification society rules. The Rules also specify the height of coamings, air pipes and door sills above the weather deck, those at the fore end being higher than the remainder.

It is also important to remove the water from the deck quickly when a heavy sea is shipped. With completely open decks, the reserve buoyancy is sufficient to lift the ship and remove the water easily. When bulwarks are fitted, however, they tend to hold back the water and this may prove dangerous. For this reason openings known as freeing ports are cut in the bulwarks, the area of the freeing ports depending upon the length of the bulwark, and again the area required for releasing the water from the deck is set out in the load line regulations.

If the freeing ports are wide, grids must be fitted to prevent crew being washed overboard. In addition, scuppers are fitted to remove the surplus water from the deck. The scuppers on the weather deck are led overboard while those on intermediate decks may be led to the bilges or, if automatic non-return valves are fitted, may be led overboard.

Type A ships, with their smaller freeboard, are more likely to have water on the deck and it is a condition of assignment that open rails be fitted instead of bulwarks.

On the older ships with the accommodation midships, a longitudinal gangway was fitted to allow passage between the after end and midships and between the forward end and midships, without setting foot on the weather deck. In larger ships it is necessary to fit shelters along the gangway. Alternatively, access may be provided by an underdeck passage, but while convenient for bulk carriers and container ships would prove dangerous in oil tankers.

Tonnage

Early rules

Gross registered tonnage came about in 1854, but in 1967 the Tonnage Rules were completely revised in an attempt to improve the safety of dry cargo ships. *A registered tonne* represents 100 cubic feet of volume and the *tonnage deck* was the second deck except in single deck ships. The *tonnage length* was measured at the level of the tonnage deck where an imaginary line was drawn inside the hold frames or sparring, the tonnage length being measured on the centreline of the ship to this line.

Tonnage depths were measured from the top of the tank or ceiling to the underside of the tonnage deck at the centreline, less one-third of the camber. There was, however, a limitation on the height of the double bottom considered. *Tonnage breadths* are measured to the inside of the hold frames or sparring.

The tonnage length was divided into a number of parts. At each cross-section the tonnage depth is similarly divided and tonnage breadths measured. The breadths are put through Simpson's Rule to give cross-sectional areas. The cross-sectional areas were then put through the Simpson's Rule to give a volume. This volume, divided by 100, is the *underdeck tonnage*.

The *gross tonnage* was found by adding to the underdeck tonnage, the tonnage of all enclosed spaces between the upper deck and the second deck, the tonnage of all enclosed spaces above the upper deck together with any portion of hatchways exceeding $\frac{1}{2}$% of the gross tonnage.

The *net tonnage or register tonnage* was then obtained by deducting from the gross tonnage, the tonnage of spaces which are required for the safe working of the ship:

- master's accommodation
- crew accommodation and an allowance for provision stores
- wheelhouse, chartroom, radio room and navigation aids room
- chain locker, steering gear space, anchor gear and capstan space
- space for safety equipment and batteries
- workshops and storerooms for pumpmen, electricians, carpenter, boatswains and the lamp room
- auxiliary/emergency diesel engine and/or auxiliary boiler space if outside the engine room
- pump room if outside the engine room
- in sailing ships, the storage space required for the sails, with an upper limit of $2\frac{1}{2}$ % of the gross tonnage
- water ballast spaces if used only for that purpose (the total deduction for water ballast, including double bottom spaces, may not exceed 19% of the gross tonnage), and
- *propelling power allowance* – this forms the largest deduction and is calculated as follows:

If the machinery space tonnage is between 13% and 20% of the gross tonnage, the propelling power allowance is 32% of the gross tonnage. If the machinery space tonnage is less than 13% of the gross tonnage, the propelling power allowance is a proportion of 32% of the gross tonnage.

Modified tonnage

Many ships are designed to run in service at a loaded draught which is much less than that allowed by the Load Line Rules. If the freeboard of a vessel is greater than that which would be assigned taking the second deck as the freeboard deck, then reduced gross and net tonnages used to be allowed. In this case the tonnage of the space between the upper deck and the second deck is not added to the underdeck tonnage and is therefore not included in the gross tonnage or net tonnage, both of which are consequently considerably reduced. As an indication that this *modified tonnage* has been allocated to the ship, a *tonnage mark* must be cut in each side of the ship in line with the deepest loadline and 540 mm aft of the centre of the disc.

If any cargo is carried in the 'tween decks, it is classed as deck cargo and added to the tonnage.

Alternative tonnage

The owner may, if he wishes, have assigned to any ship reduced gross tonnage and net tonnages calculated by the method given above. This is an *alternative* to the normal tonnages and is penalised by a reduction in the maximum draught. A *tonnage mark* must be cut in each side of the ship at a distance below the second deck depending upon the ratio of the tonnage length to the depth of the second deck. If the ship floats at a draught at or below the apex of the triangle, then the reduced tonnages may be used. If, however, the tonnage mark is submerged, then the normal tonnages must be used.

The principle behind the modified and alternative tonnages is that reduced tonnages were previously assigned if a tonnage hatch were fitted. This hatch seriously impaired the safety of the ship. Thus, by omitting the hatch the ship is more seaworthy and no tonnage penalty is incurred. The tonnage mark suitable for alternative tonnage is shown in Figure 10.1. The distance W is $\frac{1}{4B}$ × the moulded draught to the tonnage mark.

Current rules

The 1967 and earlier Tonnage Rules influenced the design of ships and introduced features which were not necessarily consistent with the safety and efficiency of the ship. In 1969 an International Convention on Tonnage Measurement of ships was held and new Tonnage Rules were produced. These Rules came into force in 1982 for new ships, although the 1967 Rules could still be applied to existing ships until 1994.

The principle behind the new Rules was to produce similar values to the previous rules for gross and net tonnage using a simplified method which reflected more closely the actual size of the ship and its earning capacity without influencing the design and safety of the ship.

The gross tonnage is calculated from the formula

Gross tonnage (gt) = $K_1 V$,

where V = total volume of all enclosed spaces in the ship in m³ and $K_1 = 0.2 + 0.02 \log_{10} V$. Thus the gross tonnage depends upon the total volume of the ship and therefore represents the size of the ship.

Enclosed spaces represent all those spaces which are bounded by the ship's hull, by fixed or portable partitions or bulkheads and by decks or coverings other than awnings.

Spaces excluded from measurement are those at the sides and ends of erections which cannot be closed to the weather and are not fitted with shelves or other cargo fitments.

The net tonnage is calculated from the formula

$$\text{Net tonnage (nt)} = K_2 V_C \left(\frac{4d}{3D}\right)^2 + K_3 \left(N_1 + \frac{N_2}{10}\right)$$

where V_C = total volume of cargo spaces in m³,

$K_2 = 0.2 + 0.02 \log_{10} V_C,$

$K_3 = 1.25 \left(\frac{gt + 10\ 000}{10\ 000}\right),$

D = moulded depth amidships in m,

d = moulded draught amidships in m,

N_1 = number of passengers in cabins with not more than 8 berths, and

N_2 = number of other passengers.

When a ship is designed to carry fewer than 13 passengers, the second term in the equation is ignored and the net tonnage is then based directly on the cargo capacity.

There are three further conditions:

The term $\left(\frac{4d}{3D}\right)$ is not to be taken as greater than unity.

The term $K_2 V_C \left(\frac{4d}{3D}\right)^2$ is not to be taken as less than 0.25 gt.

The net tonnage is not to be taken as less than 0.30 gt.

Hence ships which carry no passengers and little or no cargo will have a net tonnage of 30% of the gross tonnage. All cargo spaces are certified by permanent markings CC (cargo compartment).

The result of these rules saw the shelter deck vessels carrying dual tonnage and the tonnage mark disappear from the scene. A unified system of measurement is now used by all nations with no variations in interpretation. Net tonnage is used to determine canal dues, light dues, some pilotage dues, some harbour dues and national 'tonnage taxes'.

Gross tonnage is used to determine manning levels, safety requirements such as fire appliances, some pilotage and harbour dues, towing charges and graving dock costs.

In 2010 the IMO's International Code on Intact Stability, 2008 (2008 IS Code) came into force. This provides all the requirements for ship's designers and operators in one document.

Probabilistic method of calculating damage stability

In an effort to improve the 'survivability' of ships in the event of collision, the IMO requires designers to initially assess the internal structure of their design based upon the assumptions of damage occurring for a given type of ship, cargo carried, length, width and draught. (Called the deterministic method)

However, for passenger ships and dry-cargo ships over 80 m long designers must consider the probable survivability of the vessel in the event of different damage scenarios. The results of the survivability assessments will then require a modification to the internal structure to improve the ability of the vessel to survive the damage.

Life-Saving Appliances

The life-saving equipment carried on board a ship depends upon the number of persons carried and the normal service of the ship. A Transatlantic passenger liner would carry considerably more equipment than a coastal cargo vessel. The following notes are based on the requirements for a deep-sea cargo ship.

There must be sufficient lifeboat accommodation on *each* side of the ship for the whole of the ship's complement. The lifeboats must be at least 7.3 m long and may be constructed of wood, steel, aluminium or fibreglass. They carry rations for several days, together with survival and signalling equipment such as fishing lines, first-aid equipment, compass, lights, distress rockets and smoke flares. One lifeboat on each side must be motor-driven.

The lifeboats are suspended from davits, which allow the boats to be lowered to the water by gravity when the ship is heeled to 15°. Most modern ships are fitted with gravity davits, which, when released, allow the cradle carrying the boat to run outboard until the boat is hanging clear of the ship's side (Figure 10.2).

The boat can be raised and lowered by means of an electrically driven winch. The winch is manually controlled by a weighted lever (Figure 10.3), known as a *dead man's handle,*

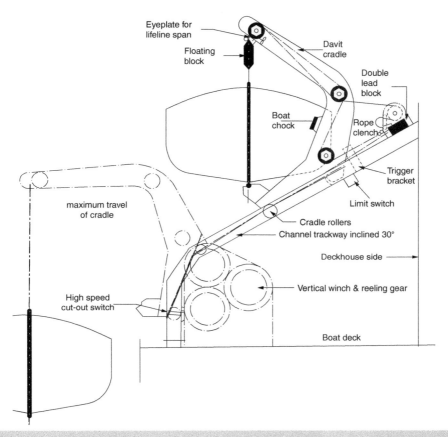

Eyeplate for lifeline span

Floating block

Davit cradle

Double lead block

Boat chock

Rope clench

Trigger bracket

maximum travel of cradle

Limit switch

Cradle rollers

Channel trackway inclined 30°

Deckhouse side

Vertical winch & reeling gear

High speed cut-out switch

Boat deck

▲ **Figure 10.2** *Gravity davit*

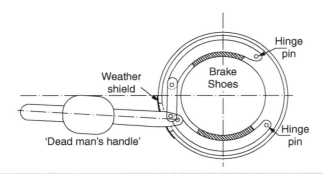

Hinge pin

Weather shield

Brake Shoes

'Dead man's handle'

Hinge pin

▲ **Figure 10.3** *Main brake*

which releases the main brake. Should the operator lose control of the brake the lever causes the winch to stop. The speed of descent is also controlled by a centrifugal brake which limits the speed to a maximum of 36 m/min. Both the centrifugal brake and the main brake drum remain stationary during the hoisting operation. If the power fails while raising the boat, the main brake will hold the boat.

Ships also carry liferafts having sufficient capacity for half of the ship's crew. The liferafts are inflatable and carry survival equipment similar to the lifeboats. They have been found extremely efficient in practice, giving adequate protection from exposure.

Each member of the crew is supplied with a lifejacket which is capable of supporting an unconscious person safely.

Lifebuoys are provided in case a man falls overboard. Some are fitted with self-igniting lights for use at night and others fitted with smoke signals for pin-pointing the position during the day. All ships carry line throwing apparatus which consists of a light line to which a rocket is attached. The rocket is fired from a pistol and must be capable of carrying 230 m. This enables contact to be made between the ship and the shore or another ship. A hawser or whip is attached to the end of the line and pulled back onto the ship either directly or through a block, allowing persons to be transferred or vessels to be towed.

With the introduction of totally enclosed lifeboats came the need to use on-load release gear which was activated from inside an enclosed boat. These devices need very special attention as the original designs did not 'fail safe'. However, after January 2015 all ships must be fitted with a new 'fail safe' design, at the first available dry-dock.

The recent development of *free fall lifeboats* has been applied to smaller vessels where the drop to the water is limited. Some manufacturers state that these boats can be built to fall from up to 40 m, however each boat will come with a certificate from the manufacturers stating the 'greatest launching height'.

Under strict control a limited number of vessels are fitted with *maritime escape systems*. These are inflatable slides or cylindrical shoots usually fitted to a limited number of passenger ferries. They are designed to evacuate a large number of passengers to inflatable liferafts.

Fire Protection

Definitions

A *non-combustible material* is one which neither burns nor gives off flammable vapours in sufficient quantity for self-ignition when heated to 750°C.

A *standard fire test* is carried out on a stiffened panel of material 4.65 m² in area, 2.44 m high with one joint. One side of the panel is exposed in a test furnace to a series of temperatures, 538°C, 704°C, 843°C and 927°C at the end of 5, 10, 30 and 60 min periods, respectively.

An *A-class division* must be made of steel or equivalent material capable of preventing the passage of smoke and flame to the end of the 60-min standard fire test. The average temperature on the unexposed surface of the panel must not rise more than 139°C after a given time and it may be necessary to insulate the material. The time intervals are 0, 15, 30 and 60 min and the divisions classed to indicate this interval, that is, A-0; A-60.

A *B-class division* need not be made of steel but must be of non-combustible material and must prevent the passage of smoke and flame to the end of the first 30 min of the standard fire test. The average temperature on the unexposed surface must not rise more than 140°C after 0 or 15 min.

A *C-class division* must be of non-combustible material.

Control stations refer to spaces containing main navigation or radio equipment, central fire-recording system or the emergency generator.

Fire potential is the likelihood of fire starting or spreading in a compartment. If the fire potential is high, then a high standard of insulation is required. Thus bulkheads separating accommodation from machinery spaces would be required to be A-60 while those dividing accommodation from sanitary spaces could be B-0 or even C.

There are three basic principles of fire protection:

1. The separation of accommodation spaces from the rest of the ship by thermal and structural boundaries.

2. The containment, automatic extinction or detection of fire in the space of origin, together with a fire alarm system.

3. The protection of a means of escape.

Passenger ships

Passenger ships are divided into main vertical zones by A-class divisions not more than 40 m apart. These divisions are carried through the main hull, superstructure and deckhouses. If it is necessary to step the bulkhead, then the deck within the step must also be A-class.

The remainder of the bulkheads and decks within the main vertical zones are A, B or C class depending upon the fire potential and relative importance of the adjacent compartments. Thus bulkheads between control stations or machinery spaces and accommodation will be A-60 while those between accommodation and sanitary spaces will be B-0.

The containment of fire vertically is extremely important and the standard of protection afforded by the decks is similar to that of the bulkheads. If a sprinkler system is fitted the standard of the division may be reduced, typically from A-60 to A-15.

All compartments in the accommodation, service spaces and control stations are fitted with automatic fire alarm and detection systems.

All stairways are of steel or equivalent and are within enclosures formed by A-class divisions. Lift trunks are designed to prevent the passage of smoke and flame between decks and to reduce draughts. Ventilator trunks and ducts passing through main vertical zone bulkheads are fitted with dampers capable of being operated from both sides of the bulkheads.

Fire resisting doors may be fitted in the A-class bulkheads forming the main vertical zone and those enclosing the stairways. They are usually held in the open position but close automatically when released from a control station or at the door position even if the ship is heeled ±3.5°.

Vehicle spaces in ships having drive-on/drive-off facilities present particular problems because of their high fire potential and the difficulty of fitting A-class divisions. A high standard of fire extinguishing is provided by means of a *drencher* system. This comprises a series of full-bore nozzles giving an even distribution of water of between 3.5 and 5.0 l/m²min over the full area of the vehicle deck. Separate pumps are provided for the system.

Dry cargo ships

In ships over 4000 tonnes gross all the corridor bulkheads in the accommodation are of steel or B-class. The deck coverings inside accommodation which lies above machinery or cargo spaces must not readily ignite. Interior stairways and crew lift trunks are of steel as are bulkheads of the emergency generator room and bulkheads separating the galley, paint store, lamp room or bosun's store from accommodation.

Oil tankers

In tankers of over 500 tonnes gross, the machinery space must lie aft of the cargo space and must be separated from it by a cofferdam or pump room. Similarly all accommodation must lie aft of the cofferdam. The parts of the exterior of the superstructure facing the cargo tanks and for 3 m aft must be A-60 standard. Any bulkhead or deck separating the accommodation from a pump room or machinery space must also be of A-60 standard.

Within the accommodation the partition bulkheads must be of at least C standard. Interior stairways and lift trunks are of steel, within an enclosure of A-0 material.

To keep deck spills away from the accommodation and service area a permanent continuous coaming 150 mm high is welded to the deck forward of the superstructure.

It is important to prevent gas entering the accommodation and engine room. In the first tier of superstructure above the upper deck no doors to accommodation or machinery spaces are allowed in the fore end and for 5 m aft. Windows are not accepted but non-opening ports may be fitted but are required to have internal steel covers. Above the first tier non-opening windows may be fitted in the house front with internal steel covers.

Classification of Ships

A classification society is an organisation whose function is to ensure that a ship is built to match its rules of construction and that the standard of construction is maintained. The ship is then classified according to the standard of construction and equipment.

The cost of insurance of both ship and cargo depends to a great extent upon this classification, and it is therefore to the advantage of the shipowner to have a high class ship. It should be noted, however, that the classification societies are independent of the insurance companies.

There are a number of large societies, each being responsible for the classification of the majority of ships built in at least one country, although in most cases it is left to the shipowner to choose the society. The top ten organisations are as follows:

Lloyd's Register of Shipping (LR)	United Kingdom
American Bureau of Shipping (ABS)	USA
Bureau Veritas (BV)	France
Det Norske Veritas, Germanischer Lloyd (DNV)	Norway/Germany
Registro Italiano (RINA)	Italy
Teikoku Kaiji Kyokai (Class NK)	Japan
China Classification Society (China Class)	China
Russian Maritime Register of Shipping (RS Class)	Russia
Indian Register of Shipping (IR)	India
Korean Register of Shipping (KR)	Korea

Each of these societies has its own rules which may be used to determine the design, composition and therefore the ultimate strength of the structural members. The following notes are based on Lloyd's Rules. Steel ships which are built in accordance with the Society's Rules, or are regarded by Lloyd's as equivalent in strength, are assigned a class in the Register Book. This class applies as long as the ships are found under survey to be in a fit and efficient condition. Class 100A is assigned to ships which are built in accordance with the rules or are of equivalent strength.

The number 1 is added (i.e. 100A 1) when the equipment, consisting of anchors, cables, mooring ropes and towropes, is in good and efficient condition. The distinguishing mark ✠ is given when a ship is fully built under Special Survey, that is, when a surveyor is in attendance and examines the ship during all stages of construction. Thus a ship classed as ✠ 100A 1 is built to the highest standard assigned by Lloyd's. Additional notations are added to suit particular types of ship such as 100A 1 oil tanker or 100A 1 ore carrier.

When the machinery is constructed and installed in accordance with Lloyd's Rules a notation LMC is assigned, indicating that the ship has Lloyd's Machinery Certificate.

In order to claim the 100A 1 class, the materials used in the construction of the ship must be of good quality and free from defects. To ensure that this quality is obtained, samples of material are tested at regular intervals by Lloyd's Surveyors.

To ensure that the ship remains worthy of its classification, annual and special surveys are carried out by the surveyors. The special surveys are carried out at intervals of 4–5 years.

In an annual survey the ship is examined externally and, if considered necessary, internally. All parts liable to corrosion and chafing are examined, together with the hatchways, closing devices and ventilators to ensure that the standards required for the Load Line Regulations are maintained. The steering gear, windlass, anchors and cables are inspected.

A more thorough examination is required at the special surveys. The shell plating, sternframe and rudder are inspected, the rudder being lifted if considered necessary. The holds, peaks, deep tanks and double bottom tanks are cleared, examined and the tanks tested. The bilges, limbers and tank top are inspected, part of the tank top ceiling being removed to examine the plating. With respect to any corroded parts, the thickness of the plating must be determined by ultrasonic testing.

The scantlings of the structure are based on theory, but because a ship is a very complex structure, a 'factor of experience' is introduced. Lloyd's receives reports of all faults and failures in ships which carry their class, and on the basis of these reports, consistent faults in any particular type of ship may be studied in detail and amendments made to the rules.

For example, structural damage in some ships did lead to the introduction of longitudinal framing in the double bottom. Reports of brittle fracture have resulted in amendments to the design rules for the shell and deck of some types of ships. It is important to note that Lloyd's have the power to require owners to alter the structure of an existing ship if they consider that the structure is weak. An example of this in the past was the fitting of butt straps to some welded tankers to act as crack arrestors, the shell plating being cut to create a discontinuity in the material, and the separate plates joined together by means of the butt strap.

In order for a ship to receive Lloyd's highest classification, scantling plans are drawn. On these plans the thicknesses of all plating, sizes of beams and girders and the method of construction are shown. The scantlings are obtained from Lloyd's Rules and depend upon the length, breadth, draught, depth and frame spacing of the ship and the span of the members. Variations may occur due to special design characteristics such as the size and position of the machinery space. Shipowners are at liberty to increase any of the scantlings and many do so, particularly where such increases lead to reduced repair costs. An increase in diameter of rudder stock by 10% above rule is a popular owners' extra. Shipbuilders may also submit alternative arrangements to those given in the rules and Lloyd's may allow their use if they are equivalent in strength. The scantling plans are submitted to Lloyd's for their approval before detailed plans are drawn and the material ordered.

The number of *deep water berths* is on the increase as the size of oil tankers, bulk carriers and container ships continues to grow. This has led to an increase of the system of hull survey while the vessel is afloat. The in-water survey (IWS) is used to check those parts of a ship which are usually surveyed in dry dock. It includes visual examination of the hull, rudder, propeller, sea inlets and so on, and the measurement of wear down of rudder bearings and stern bush.

There are several requirements to be met before IWS is allowed. High resolution colour photographs are taken of all parts which are likely to be inspected, before the ship is launched. The rudder and sternframe are designed for easy access to bearings. The ship must be less than 10 years old, have a high resistance coating on the underwater hull and be fitted with an impressed current cathodic protection system.

At the time of the inspection the hull is cleaned by one of the many brush systems available. The water must be clear and the draught less than 10 m. The inspection may be made by an underwater closed-circuit television camera. The camera may be hand-held by a diver or carried by a hydraulically propelled camera vehicle, remotely controlled from a surface monitoring station. The use of remotely operated unmanned vehicles is transforming this area of ship inspection/maintenance.

Survey of Ships – Discontinuities

If there is an abrupt change in section in any type of stressed structure, particularly high stresses occur at the discontinuity. Should the structure be subject to fluctuating loads, the likelihood of failure at this point is greatly increased.

A ship is a structure in which discontinuities are impossible to avoid. It is also subject to fluctuations or even reversals of stress when passing through waves. The ship must be designed to reduce such discontinuities to a minimum, while great care must be taken in the design of structural detail with respect to any remaining changes in section.

The most highly stressed part of the ship structure is usually that within 40%–50% of its length amidships. Within this region every effort must be made to maintain a continuous flow of material.

Difficulties occur at hatch corners (Figure 10.4). Square corners must be avoided and the corners should be radiused or elliptical. With radiused corners the plating at the corners must be thicker than the remaining deck plating. Elliptical corners are more efficient in reducing the corner stress and no increase in thickness is required.

Similarly openings for doors, windows, access hatches, ladderways and so on in all parts of the ship must have rounded corners with the free edges dressed smooth.

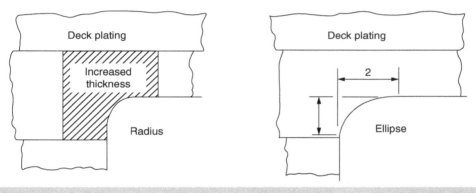

▲ **Figure 10.4** *Hatch corners*

If a bridge structure is fitted over more than 15% of the ship's length, the bridge side plating must be tapered or curved down to the level of the upper deck. The sheerstrake is increased in thickness by 50% and the upper deck stringer plate by 25% at the ends of the bridge. Four 'tween deck frames are carried through the upper deck into the bridge space at each end of the bridge to ensure that the ends are securely tied to the remaining structure.

In bulk carriers, where a large proportion of the deck area is cut away to form hatches, the hatch coamings should preferably be continuous and tapered down to the deck level at the ends of the ship.

Longitudinal framing must be continued as far as possible into the ends of the ship and scarphed gradually into a transverse framing system (Figure 10.5). This problem is overcome in oil tankers and bulk carriers by carrying the deck and upper side longitudinals through to the collision bulkhead, with transverse framing in the fore

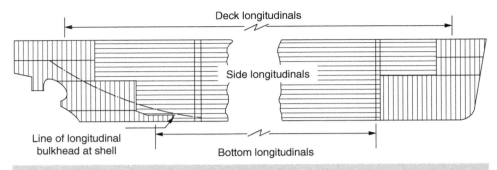

▲ **Figure 10.5** *Extent of longitudinal framing oil tanker*

deep tank up to the tank top level and in the fore peak up to the upper deck. Similarly at the after end the side and deck longitudinals are carried aft as far as they will conveniently go. A particular difficulty arises with the longitudinal bulkheads which must be tapered off into the forward deep tank and engine room. In the larger vessels it is often possible to carry the bulkheads through the engine room, the space at the sides being used for auxiliary spaces, stores and workshops.

A similar problem occurs in container ships, when it is essential to taper the torsion box and similar longitudinal stiffening gradually into the engine room and the fore end. The fine lines of these ships cause complications which are not found in the fuller vessels. In these faster vessels the importance of tying the main hull structure efficiently into the engine room structure cannot be sufficiently emphasised.

Load line surveys

To ensure that the vessel is maintained at the same standard of safety as when it was built, annual surveys are made by the Assigning Authority. An inspection is made of all those items which affect the freeboard of the ship and are included in the Conditions of Assignment. A note is made about of any alterations to the ship which could affect the assigned freeboard.

The main areas of the ship that are checked relate to the:

- visibility of the freeboard marks on both sides of the vessel
- accuracy of the freeboard marks
- quality of the stability information available to the officers

- condition of any openings that need to be watertight
- ability of the vessel to allow water to drain from the decks
- structure is of sufficient strength – no significant corrosion or damage which would impact on the design strength of the vessel
- suitability of the crew accommodation.

11

SHIP DYNAMICS

Propellers

This section should be read in conjunction with *Reeds Vol 4*, Chapter 8.

The design of a propulsion system for a ship is required to be efficient for the ship in its intended service, reliable in operation, free from vibration and cavitation and economical in initial cost, running costs and maintenance. Some of these factors conflict with others and, as with many areas of engineering, the final system is a compromise. Various options are open to the shipowner including the number of blades, the number of propellers, the type and design of propellers and shaft line or electric drive.

Propellers work in an adverse environment created by the varying wake field produced by the after end of the ship at the propeller disc. Figure 11.1 shows a typical wake distribution for a single screw ship. The localised calculation of the wake fraction W_T is determined by wake speed ÷ ship speed.

High wake fractions indicate that the water is being carried along at almost the same speed as the ship. Thus the propeller is working in almost *dead* water. The lower fractions indicate that the water is almost stationary and therefore has a high speed relative to the propeller. As the propeller blade passes through these different regions it is subject to a fluctuating load. These variations in loading cause several problems.

For example, consider a four-bladed propeller turning in the wake field shown in Figure 11.1, two of the blades are lightly loaded and two heavily loaded when the blades are in the position shown in the figure. When the propeller turns through 90° the situation is reversed. The resulting fluctuations in stress may produce cracks at the root of the blades and vibration of the blades.

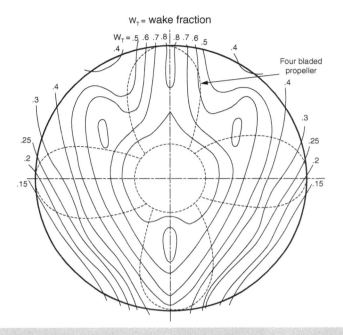

▲ **Figure 11.1** *Typical wake distribution – single screw ship*

The fluctuating loads might be reduced by changing the number of the blades or cleaning up the wake field. A three-bladed propeller, for example, will have only one blade fully loaded or lightly loaded at any one time, while five, six and seven blades produce more gradual changes in thrust and torque per blade and hence reduce the possibility of vibration due to this cause.

An alternative method of reducing the variation in blade loading is to fit a *skewed* propeller (Figure 11.2) in which the centreline of each blade is curved to spread the distribution of the blade area over a greater range of wake contours. In these propellers there is also less cavitation produced and under some conditions there are efficiency gains as well as a reduction in vibration.

Fluctuating Forces Caused by the Propeller Wake Field

The thrust of a propeller depends upon the acceleration of a mass of water within its own boundary of influence. If a propeller is on the centreline of a ship, it lies within

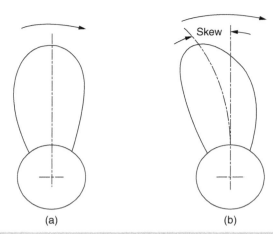

▲ **Figure 11.2** *(a) Normal blade and (b) skewed blade*

the wake field moving past the hull and therefore it accelerates water which is already moving. The disturbed water moving towards the blade will create a fluctuating pressure entering the propeller disc as it rotates (Figure 11.1). The fluctuating pressure will set up vibration and also differencing effectiveness in the thrust produced across the propeller blade.

A propeller which is off the centreline lies only partly within the wake field and therefore has a wider variation in pressure differential to contend with as some of the blade is working in slower moving water which is outside of the influence of the hull. For this reason, single screw ships can be slightly more efficient than twin screw ships for similar conditions.

The main advantages of twin screw ships are their increased manoeuvrability and the duplication of propulsion systems leading to improved safety. Set against this is the considerable increase in the cost of the construction of the after end, whether the shaft support is by A-frames or by spectacle frames and bossings, compared with the sternframe of a single screw ship.

In a single screw ship, the rudder is also more effective since it lies directly in the outflow from the propeller and hence the velocity of water at the rudder is increased, producing increased rudder force. Conversely, many twin screw ships are fitted with twin rudders in line with the propellers, further increasing their manoeuvrability at the expense of increased cost of steering gear. The variation in wake in a twin screw ship can be less than with a single screw ship, due to the propellers working in smoother water, and the blades are therefore less liable to vibrate due to fluctuations in thrust. The support of the shafting and propeller is less rigid, however, and vibration may occur due to the deflection of the support. The most recent developments in ship design have concentrated on cleaning up the 'wake field' and if this is even with much less pressure differentials then the vibration problems disappear.

Page 134 describes the Mewis duct which accomplishes this task. The development of podded drives also helps as the pod's propeller is presented to a relatively undisturbed wake field.

Podded Drive

With the developments in diesel electric propulsion systems (*Reeds Vol 12,* Chapter 9), the efficiency of the propeller has taken a jump forward.

The major reason for this is that the pods can be placed in the most beneficial position under the hull. The pods usually face forward taking advantage of the less disturbed water in this position, thus reducing pressure fluctuations as the propeller(s) rotate.

Vessels fitted with pods do not need rudders or stern thrusters, as the units are arranged so that they are able to turn 360°, which makes the design of the aft end of the vessel much cleaner and the hull presents much less resistance to movement resulting in a much more efficient arrangement overall.

The developments in control technology mean that the podded drive can be arranged as a variable speed motor driving a fixed pitch propeller. This means that the propeller will be efficient at full sea speed, as the blade area of the propeller is maximized for a given size of propeller. This is because it does not need to be a Controllable Pitch Propeller (CPP). See below for further details about the differences in propeller design.

Controllable Pitch Propellers

A controllable pitch (c.p.) propeller is one that always rotates in the same direction but the pitch of the blades may be altered by remote control. The blades are separately mounted onto bearing rings in the propeller hub. A valve rod is fitted within the hollow main shaft and this actuates a servo-motor cylinder. Longitudinal movement of this cylinder transmits a load through a crank pin and sliding shoe to rotate the propeller blade (Figure 11.3).

The propeller pitch is controlled directly from the bridge and hence closer and quicker control of the ship speed is obtained. This is of particular importance when

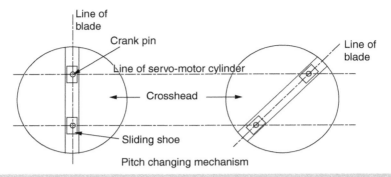

▲ **Figure 11.3** *Pitch changing mechanism*

manoeuvring in confined waters when the ship speed may be changed, and indeed reversed, at constant engine speed. Because full power may be developed astern, the stopping time and distance may be considerably reduced.

The initial cost of a c.p. propeller is considerably greater than that of a fixed pitch installation. On the other hand a simpler non-reversing main engine may be used or a reversing gear box is not required, and since the engine speed may be maintained at all times the c.p. installation lends itself to the fitting of shaft-driven auxiliaries such as a shaft alternator. The efficiency of a c. p. propeller is less than that of a fixed pitch propeller for optimum conditions, due to the larger diameter of boss required, but at different speeds the c.p. propeller has the advantage. The cost of repair and maintenance is high compared with a fixed pitch propeller although it might be possible to repair or replace a single blade of the c.p. arrangement.

Contra-Rotating Propellers

This system consists of two propellers in line, but turning in opposite directions. The after propeller is driven by a normal solid shaft. The forward propeller is driven by a short hollow shaft which encloses the solid shaft. The forward propeller is usually larger and has a different number of blades from the after propeller to reduce the possibility of vibration due to blade interference.

Research has shown that the system may increase the propulsion efficiency by 10%–12% by cancelling out the rotational losses imparted to the stream of water passing through the propeller disc. Contra-rotating propellers are extremely costly and are suitable only for highly loaded propellers and large single screw tankers, particularly when the draught is limited. The increased surface area of the combined system reduces the possibility of cavitation but the longitudinal displacement of the propellers is very critical.

Vertical Axis Propellers

The Voith–Schneider propeller is typical of a vertical axis propeller and consists of a series of vertical blades set into a horizontal rotor which rotates about a vertical axis. The rotor is flush with the bottom of the ship and the blades project down as shown in Figure 11.4.

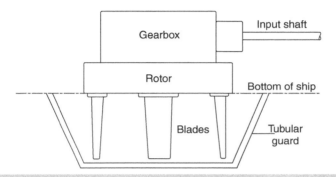

▲ **Figure 11.4** *Voith–Schneider drive*

The blades are linked to a control point P by cranked control rods (Figure 11.5). When P is in the centre of the disc, the blades rotate without producing a thrust. When P is moved away from the centre in any direction, the blades turn in the rotor out of line with the blade orbit and a thrust is produced. The direction and magnitude of the thrust depends upon the position of P. Since P can move in any direction within its inner circle, the ship may be driven in any direction and at varying speeds.

Thus the Voith–Schneider propeller may propel and manoeuvre a ship without the use of a rudder.

The efficiency of a Voith–Schneider propeller is relatively low but it has the advantage of high manoeuvrability and is useful in harbour craft and ferries. Two or more installations may be fitted and in special vessels (e.g. firefloats) can move the ship sideways or rotate it in its own length.

Replacement of damaged blades is simple although they are fairly susceptible to damage. A tubular guard is usually fitted to protect the blades. The propeller may be driven by a vertical axis motor seated on the top or by a diesel engine with a horizontal shaft converted into vertical drive by a bevel gear unit.

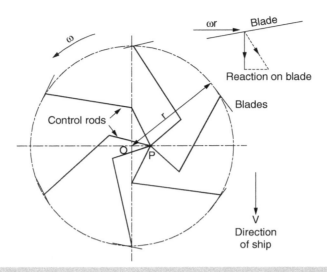

▲ **Figure 11.5** *Blade positions*

Tunnel Thrusters (Bow and Stern Thrusters)

Many ships are fitted with bow thrust units to improve their manoeuvrability (Figure 11.6). They are an obvious feature in ships working within, or constantly in and out of harbour where close control is obtained without the use of tugs. They have also proved to be of considerable benefit to larger vessels such as oil tankers and bulk carriers, where the tug requirement has been reduced.

Several types of tunnel thrusters are available, each having its own advantages and disadvantages.

In all cases the necessity to penetrate the hull forward causes an increase in ship resistance and hence in fuel costs, although the increase is small and with the podded drives the tunnel thrusters at the stern of the vessel are not required.

A popular arrangement is to have a cylindrical duct passing through the ship from side to side, in which is fitted an impeller that can produce a thrust to port or to starboard. The complete duct must lie below the waterline at all draughts, the impeller acting best when subject to a reasonable head of water and thus reducing the possibility of cavitation.

The impeller may be of fixed pitch with a variable-speed motor which is reversible or has reverse gearing. Alternatively a controllable pitch impeller may be used, having a constant-speed drive. Power may be provided by an electric motor, a diesel engine or a hydraulic motor.

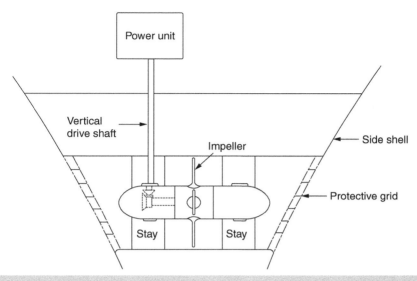

▲ **Figure 11.6** *Controllable pitch bow thrusters*

Some vessels are fitted with Voith–Schneider propellers within the ducts to produce the transverse thrust.

As an alternative the water may be drawn from below the ship and projected port or starboard through a horizontal duct which may lie above or below the waterline. A uni-directional horizontal impeller is fitted in a vertical duct below the waterline. The lower end of the duct is open to the sea, while the upper end leads into the horizontal duct which has outlets in the side shell port and starboard.

Within this duct two hydraulically operated vertical vanes are fitted to each side (Figure 11.7). Water is drawn from the bottom of the ship into the horizontal duct. By varying the position of the vanes the water jet is deflected either port or starboard, producing a thrust and creating a reaction which pushes the bow in the opposite direction.

This system has an advantage that by turning all vanes 45° either forward or aft an additional thrust forward or aft can be produced. A forward thrust would act as a retarding force while an aft thrust would increase the speed of the ship. These actions may be extremely useful in handling a ship in congested harbours. The efficiency of propeller thruster falls off rapidly as the ship speed increases. The rudder thrust, on the other hand, increases in proportion to the square of the ship speed, being

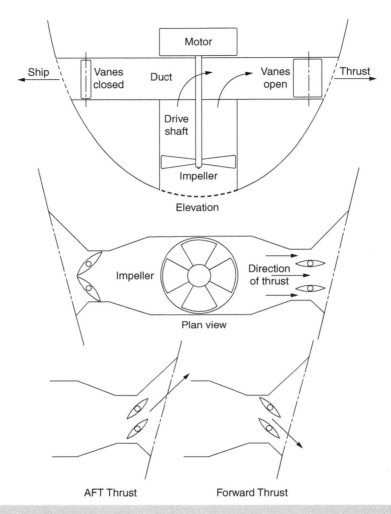

▲ **Figure 11.7** *Hydraulic thrust unit*

relatively ineffective at low speeds. The water jet unit appears to maintain its efficiency at all speeds, although neither type of thrust unit would normally be used at speed. Figure 11.8 does indicate the usefulness of thrust units when moving and docking compared with the use of the rudder.

Rolling and Stabilisation

When a ship is heeled by an external force, and the force is suddenly removed, the vessel will roll to port and starboard with a rolling period which is almost constant. This

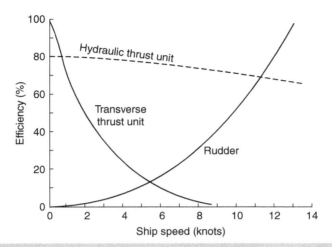

▲ **Figure 11.8** *Comparison of efficiencies*

▲ **Figure 11.9** *Typical damping curve*

is known as the ship's *natural rolling period*. The *amplitude* of the roll will depend upon the applied heeling moment and the stability of the ship. For angles of heel up to about 15° the rolling period does not vary with the angle of roll. The angle reduces slightly at the end of each swing and will eventually dampen out completely. This dampening is caused by the frictional resistance between the hull and the water, which causes a mass of *entrained water* to move with the ship (Figure 11.9).

The natural rolling period of a ship may be estimated by the formula:

$$\text{Rolling period } P = \frac{2\pi k}{\sqrt{g \text{ GM}}} \text{ seconds,}$$

where GM is the metacentric height, and k is the radius of gyration of the loaded ship about a longitudinal polar axis.

Thus a large metacentric height will produce a small period of roll, although the movement of the ship may be decidedly uncomfortable and possibly dangerous. A small metacentric height will produce a long period of roll and smooth movement of the ship. The resistance to heel, however, will be small and consequently large amplitudes of roll may be experienced. The value of the radius of gyration will vary with the disposition of the cargo. For dry cargo ships, where the cargo is stowed right across the ship, the radius of gyration varies only slightly with the condition of loading and is about 35% of the midship beam.

It is difficult in this type of ship to alter the radius of gyration sufficiently to cause any significant change in the rolling period. Variation in the period due to changes in metacentric height are easier to achieve.

In tankers and bulk carriers vessels it is possible to change the radius of gyration and not as easy to change the metacentric height. If the cargo is concentrated in the centre compartments, with the wing tanks empty, the value of the radius of gyration is small, producing a small period of roll. If, however, the cargo is concentrated in the wing compartments, the radius of gyration increases, producing a slow rolling period. This phenomenon is similar to a skater spinning on ice; as the arms are outstretched the spin is seen to be much slower.

Problems may occur in a ship which travels in a beam sea, if the period of encounter of the waves synchronises with the natural frequency of roll. Even with small wave forces the amplitude of the roll may increase to alarming proportions. In such circumstances it may be necessary to change the ship's heading and alter the period of encounter of the waves.

Reduction of Roll

Bilge keels

This plating projects from the hull and are arranged at the bilge to lie above the line of the bottom shell and within the breadth of the ship, thus being partially protected against damage. The depth of the bilge keels depends to some extent on the size of the ship, but there are two main factors to be considered:

1. The web must be deep enough to penetrate the boundary layer of water travelling with the ship.
2. If the web is too deep, the force of water when rolling may cause damage.

Bilge keels 250–400 mm in depth are fitted to ocean-going ships. The keels extend for about one-half of the length of the ship and take in the midships section. They should be continuous and are tapered gradually at the ends with the ends terminated on an internal stiffening member. Two forms of bilge keel are shown in Figure 11.10 and they are usually fitted in two parts, the connection to the shell plating being stronger than the connection between the two parts. In this way it is more likely, in the event of damage, that the web will be torn from the connecting angle rather than the connecting angle from the shell plating.

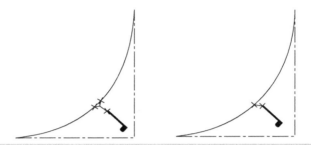

▲ **Figure 11.10** *Bilge keels*

The bilge keels reduce the initial amplitude of roll as well as subsequent movements.

Active fin stabilisers

Two fins extend from the ship side at about bilge level. They are turned in opposite directions as the ship rolls. The forward motion of the ship creates a force on each fin and hence produces a moment opposing the roll. When the fin is turned down, the water exerts an upward force. When the fin is turned up, the water exerts a downward force (Figure 11.11).

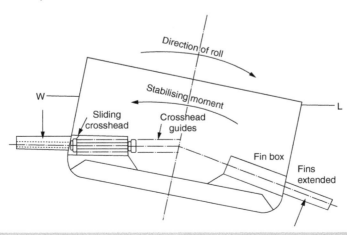

▲ **Figure 11.11** *Fin stabiliser*

The fins are usually rectangular, having aerofoil cross-section (Figure 11.12) and turn through about 20°. Many are fitted with tail fins which turn relative to the main fin through a further 10°. The fins are turned by means of an electric motor driving a variable delivery pump, delivering oil under pressure to the fin tilting gear. The oil actuates rams coupled through a lever to the fin shaft.

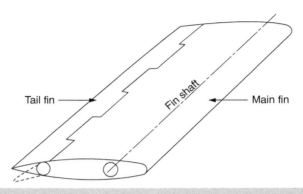

▲ **Figure 11.12** *Stabiliser fin*

Most fins are retractable, either sliding into fin boxes transversely or hinged into the ship. Hinged fins are used when there is a restriction on the width of ship, which may be allocated, such as in a container ship (Figure 11.13).

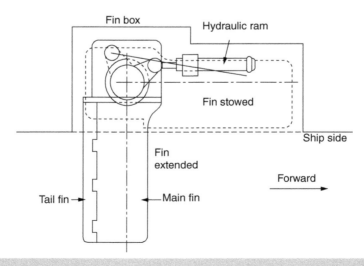

▲ **Figure 11.13** *Hinged fin*

The equipment is controlled by means of two gyroscopes, one measuring the angle roll and the other the velocity of roll. The movements of the gyroscopes actuate relays which control the angle and direction through which the fins are turned. It should be noted that no movement of stabiliser can take place until there is an initial roll of the ship and that the fins require a forward movement of the ship to produce a righting moment.

Tank stabilisers

There are three basic systems of roll-damping using free surface tanks:

1. Passive tanks.
2. Controlled passive tanks.
3. Active controlled tanks.

These systems do not depend upon the forward movement of the ship and are therefore suitable for vessels such as drill ships. In introducing a free surface to the ship, however, there is a reduction in stability which must be considered when loading the ship.

Passive tanks

Two wing tanks are connected by a duct having a system of baffles (Figure 11.14). The tanks are partly filled with water. When the ship rolls, the water moves across the system in the direction of the roll. As the ship reaches its maximum angle and commences to return, the water, slowed by the baffles, continues to move in the same direction. Thus a moment is created, reducing the momentum of the ship and hence the angle of the subsequent roll (Figure 11.15).

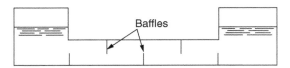

▲ **Figure 11.14** *Passive tank system*

The depth of water in the tanks is critical and, for any given ship, depends upon the metacentric height. The tank must be tuned for any loaded condition by adjusting the level, otherwise the movement of the water may synchronise with the roll of the ship

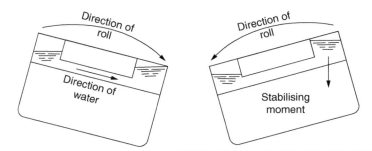

▲ **Figure 11.15** *Passive tank system in action*

and create dangerous rolling conditions. Alternatively the cross-sectional area of the duct may be adjusted by means of a gate valve.

Controlled passive tanks

The principle of action is the same as for the previous system, but the transverse movement of the water is controlled by valves operated by a control system similar to that used in the fin stabiliser. The valves may be used to restrict the flow of water in a U-tube system, or the flow of air in a fully enclosed system (Figure 11.16).

The mass of water required in the system is about 2– $2\frac{1}{2}$ % of the displacement of the ship.

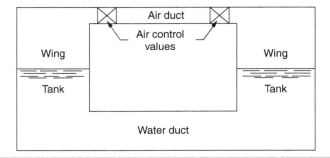

▲ **Figure 11.16** *Controlled passive tank*

Active controlled tanks

In this system the water is positively driven across the ship in opposition to the roll. The direction of roll, and hence the required direction of the water, changes rapidly. It is therefore necessary to use a uni-directional impeller in conjunction with a series of valves. The impeller runs continually and the direction of the water is controlled by valves which are activated by a gyroscope system similar to that used for the fin stabiliser (Figure 11.17).

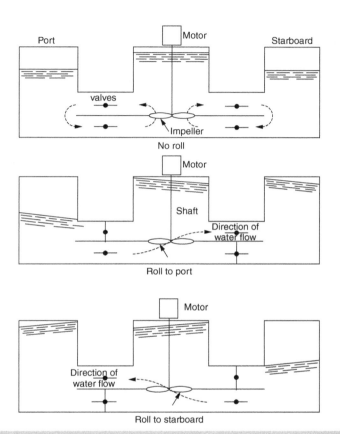

▲ **Figure 11.17** *Active controlled tank*

Vibration

Ship vibration is the periodic movement of the structure and may occur vertically, horizontally or torsionally.

There are several sources of ship vibration, any of which could cause discomfort to personnel, damage to fittings and instruments and structural failure. If the frequency of the main or auxiliary machinery at any given speed coincides with the natural frequency of the hull structure, then vibration may occur. In such circumstances it is usually easier to alter the source of the vibration by changing the engine speed or fitting dampers, than to change the structure. The natural frequency of the structure depends upon the length, mass distribution and second moment of area of the structural material. For any given mass distribution a considerable change in structural material would be required to cause any practical variation in natural frequency. There is a possibility of altering the natural frequency of the hull by redistributing the cargo. If the cargo is concentrated at the nodes, the natural frequency will be increased. If the cargo is concentrated at the anti-nodes, the natural frequency and the deflection will be reduced. Such changes in cargo distribution may only be possible in vessels such as oil tankers or bulk carriers in the ballast condition.

Similarly vibration may occur in a machinery space due to unbalanced forces from the main or auxiliary machinery or as a result of uneven power distribution in the main engine. This vibration may be transmitted through the main structure to the superstructure, causing extreme discomfort to the personnel.

As explained earlier the variation in blade loading due to the wake may create vibration of the after end, which may be reduced by changing the number of blades. The turbulence of the water caused by the shape of the after end is also a source of vibration which may be severe. It is possible to design the after end of the ship to reduce this turbulence resulting in a smoother flow of water into the propeller disc.

Severe vibration of the after end of some ships is caused by insufficient propeller tip clearance. As the blade tip passes the top of the aperture it attempts to compress the water. This creates a force on the blade which causes bending of the blade and increased torque in the shaft. The periodic nature of this force, that is, revs × number of blades, produces the vibration of the stern. Classification societies recommend minimum tip clearances to reduce propeller-induced vibrations to reasonable levels. Should the tip clearance be constant, for example, with a propeller nozzle, then this problem does not occur. If an existing ship suffers an unacceptable level of vibration from this source, it may be necessary to crop the blade tips, reducing the propeller efficiency, or to fit a propeller of smaller diameter.

A damaged propeller blade will create out-of-balance moments due to the unequal weight distribution and the uneven loading on the blades. Little may be done to relieve the resultant vibration except to repair or replace the propeller.

Wave induced vibrations may occur in ships due to pitching, heaving, slamming or the passage of waves along the ship. In smaller vessels pitching and slamming are the main sources but are soon dampened. In ships over about 300 m in length, however, hull vibration has been experienced in relatively mild sea states due to the waves. In some cases the vibration has been caused by the periodic increase and decrease in buoyancy with regular waves much shorter than the ship, while in other cases, with non-uniform waves, the internal energy of the wave is considered to be the source. Such vibrations are dampened by a combination of the hull structure, the cargo, the water friction and the generation of waves by the ship.

12

CORROSION, COATINGS AND DRY-DOCKING

Insulation of Ships

Steelwork is a good conductor of heat and is therefore said to have a high thermal conductivity. It will therefore be appreciated that some form of insulation, having low thermal conductivity, must be fitted to the inner face of the steelwork in refrigerated compartments to reduce the transfer of heat. The ideal form of insulation is a vacuum although good results may be obtained using air pockets. Most insulants are composed of materials having entrapped air cells, such as cork, glass fibre and foam plastic. Cork may be supplied in slab or granulated form, glass fibre in slab form or as loose fill, while foam plastic may either be supplied as slabs or the plastic may be foamed into position. Granulated cork and loose glass fibre depend to a large extent for their efficiency on the labour force, while in service they tend to settle, leaving voids at the top of the compartment. These voids allow increased heat transfer and plugs are fitted to allow the spaces to be repacked. Horizontal stoppers are arranged to reduce settlement. Glass fibre has the advantages of being fire resistant and vermin-proof and will not absorb moisture. Since it is also lighter than most other insulants it has proved very popular in modern vessels. Foam plastic has recently been introduced, however, and when foamed in position, entirely fills the cavity even if it is of awkward shape.

The depth of insulation in any compartment depends upon the temperature required to maintain the cargo in good condition, the insulating material and the depth to which any part of the structure penetrates the insulant. It is usually found that the depths of frames and beams govern the thickness of insulation at the shell and decks, 25–50 mm of insulation projecting past the toe of the section. It therefore proves economical in insulated ships to use frame and reverse as shown in Figure 3.7 in Chapter 3 and thus reduce the depth of insulation. This also has the effect of increasing the cargo capacity.

The internal linings required to retain and protect the insulation may be of galvanised iron, stainless steel or aluminium alloy. The linings are screwed to timber grounds which are in turn connected to the steel structure. The linings are made airtight by coating the overlaps with a composition such as white lead and fitting sealing strips. This prevents heat transfer due to a circulation of air and prevents moisture entering the insulation. Cargo battens are fitted to all the exposed surfaces to prevent contact between the cargo and the linings and to improve the circulation of air round the cargo. Figure 12.1 shows a typical arrangement of insulation at the ship side.

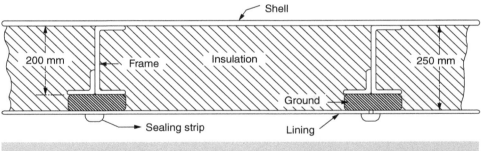

▲ **Figure 12.1** *Shell insulation*

At the tank top an additional difficulty arises – that of providing support for the cargo. Much depends upon the load bearing qualities of the insulant. Slab cork, for instance, is much superior in this respect to glass fibre, and may be expected to carry some part of the cargo load. The tank top arrangement in such a case would be as shown in Figure 12.2.

Slab cork 150–200 mm thick is laid on hot bitumen. The material is protected on its upper surface by a layer of asphalt 50 mm thick which is reinforced by steel mesh. Wood ceiling is fitted under the hatchways where damage is most likely to occur.

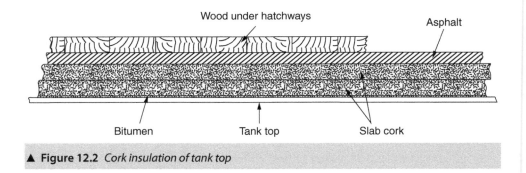

▲ **Figure 12.2** *Cork insulation of tank top*

Figure 12.3 shows the arrangement where glass fibre is used. The cargo is supported on double timber ceiling which is supported by bearers about 0.5 m apart. An additional thickness of ceiling is fitted under the hatchways. The glass fibre is packed between the bearers. If oil fuel is carried in the double bottom with respect to an insulated space, it is usual to leave an air gap between the tank top and the insulation. This ensures that any leakage of oil will not affect the insulation.

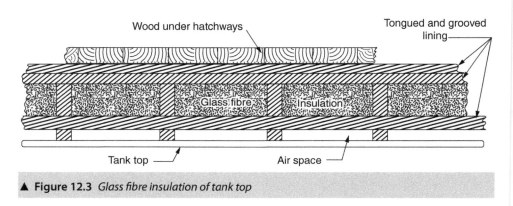

▲ **Figure 12.3** *Glass fibre insulation of tank top*

Particular care must be taken to design the hatchways to avoid heat transfer. The normal hatch beams are fitted with tapered wood which is covered with galvanised sheet. A similar type of arrangement is made at the ends of the hatch. Insulated plugs with opposing taper are wedged into the spaces. The normal hatch boards are fitted at the top of the hatch as shown in Figure 12.4. Steel hatch covers may be filled with some suitable insulation and do not then require separate plugs.

Similar types of plug are fitted at the bilge to allow inspection and maintenance, and with respect to tank top manholes.

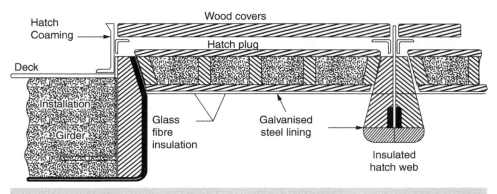

▲ **Figure 12.4** *Insulated hatch plug*

Drainage of insulated spaces is rather difficult. Normal forms of scupper would lead to increase in temperature. It becomes necessary, therefore, to fit brine traps to all 'tween deck and hold spaces. After defrosting the compartments and removing the cargo the traps must be refilled with saturated brine, thus forming an air seal which will not freeze. Figure 12.5 shows a typical brine trap fitted in the 'tween decks.

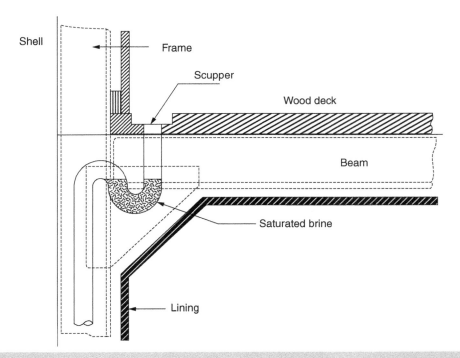

▲ **Figure 12.5** *Tween deck scupper*

Corrosion

Corrosion is the wasting away of a material due to its tendency to return to its natural state, which, in the case of a metal, is in the form of an oxide.

If a metal or alloy is left exposed to a damp atmosphere, an oxide will form on the surface. If this layer is insoluble, it forms a protective layer which prevents any further corrosion. Copper and aluminium are two such metals. If, on the other hand, the layer is soluble, as in the case of iron, the oxidation continues, together with the erosion of the material.

When two dissimilar metals or alloys are immersed in an electrolyte, an electric current flows through the liquid from one metal to the other and back through the metal pathway. The direction in which the current flows depends upon the relative position of the metals in the electrochemical series. For common metals in use in ships, the current flows from the anode to the cathode, which is higher in the Periodic table, or more electro- positive. Thus, if copper and iron are joined together and immersed in an electrolyte, a current will flow through the electrolyte from the iron to the copper and back through the copper to the iron (Figure 12.6).

Positive copper	+
Lead	
Tin	
Iron	
Chromium	
Zinc	
Aluminium	
Negative magnesium	−

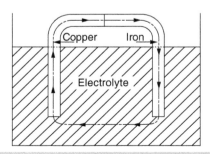

▲ **Figure 12.6** *Corrosion cell*

Unfortunately, during this process, material from the anode is transferred to the cathode, resulting in corrosion of the anode. Slight differences in potential occur in the same material. Steel plate, for instance, is not perfectly homogeneous and will therefore have anodic and cathodic areas. Corrosion may therefore occur when such a plate is immersed in an electrolyte such as sea water. The majority of the corrosion of ships is due to this electrolytic process.

Hull Coatings – for Efficiency and for the Prevention of Corrosion

When metal is subjected to moisture and oxygen the original parent metal starts to degrade in a chemical process known as corrosion. The corrosion of metal may be prevented by coating the metallic surface with a substance which prevents contact with moisture and/or oxygen. This does seem like a simple principle to achieve; in practice, however it is proved difficult to maintain such a coating, particularly on ships which are subjected to the possibility of mechanical damage.

When a vessel is travelling across the ocean the hull might come into contact with floating or submerged objects that could cause damage. While arriving at or leaving the berth the hull could be subjected to damage by coming into contact with the quay and while the ship is stationary, either at anchor or alongside, the various marine organisms like to attach themselves to the hull.

Therefore, a modern hull coating has to accomplish two major objectives, which are to stop the hull corroding and to stop the hull becoming fouled with marine growth.

Anti-fouling can be accomplished by using:

- coating containing a biocide that kills the marine growth
- 'low adhesion' coating, allowing the growth to be washed away by the moving ship (sometimes referred to as 'self-polishing')
- very hard coating possibly containing glass flakes.

Some coating systems are also designed to become smoother as the vessel moves through the water. This improves performance as the vessel progresses from one docking to the next.

Surface preparation

New building

Steel plates supplied to shipbuilders have patches of a black oxide known as *mill scale* adhering to the surface. This scale is insoluble and, if maintained over the whole surface, would reduce corrosion. It is, however, very brittle and does not expand either mechanically or thermally at the same rate as the steel plate. Unless this mill scale is removed before coating, the covered scale will drop off in service, leaving bare steel plate which will corrode rapidly. Unfortunately mill scale is difficult to remove completely.

If the plate is left exposed to the atmosphere, rust will form behind the mill scale. On wire brushing, the majority of the scale will be removed. This is known as *weathering*. In modern times a good flow of material through the shipyard is essential and therefore the time allowed for weathering must be severely limited. In addition, it is found in practice that much of the mill scale is not removed by this process.

If the plates are immersed in a weak solution of sulfuric acid or hydrochloric acid for a few hours, the majority of the mill scale is removed. This system, known as *pickling*, has been used by the Admiralty and several private owners for many years. The pickled plate must be hosed down with fresh water on removal from the tank, to remove all traces of the acid. It is then allowed to dry before painting. One disadvantage with this method is that during the drying period a light coating of rust is formed on the plate and must be removed before painting.

Flame cleaning of a ship's structure came into use some time ago. An oxy-acetylene torch, having several jets, is used to brush the surface. It burns any dirt and grease, loosens the surface rust and, due to the differential expansion between the steel and the mill scale, loosens the latter. The surface is immediately wire brushed and the priming coat applied while the plate is still warm. This method is not usually used on a large scale any more.

The most effective method of removing the mill scale to date is the use of *shot blasting*. The steel plates are passed through a machine in which steel shot is projected at the plate, removing the mill scale together with any surface rust, dirt and grease. This system removes 95–100% of the mill scale and results in a slightly rough surface which allows adequate adhesion of the paint. In modern installations, the plate is spray painted on emerging from the shot blasting machine.

These methods of preparing steel are still used; however the ship builder will require the steel to be supplied in this 'pre-prepared' condition. The steel will arrive with the mill scale removed and the steel coated with a thin coat of primer. This 'shop primer' gives temporary protection from the atmospheric conditions while it is transported and stored ready for use.

Surface preparation during dry-docking

One of the primary reasons for docking a ship is to inspect the hull for any damage to the structure, to repair the damage and to reinstate the hull coating.

Surface preparation is crucial to the performance of any coating system. It is desirable but sometimes not cost-effective to take the surface back to bare metal. This is termed Sa 2½ which comes from the Swedish Standards for the cleanliness of the metal surface. Sa 3 is totally bare metal with Sa 2½ being a practical approximation to Sa 3.

If changing from one coating system to another, then the hull should be shot/sand blasted to Sa 2½ before the new system is applied. A second reason for paying attention to the preparation of the hull is to do with the roughness that preparation can bring to the hull of a ship.

Some owners will only allow the hull to be prepared in areas where there is corrosion visible. This is known as 'spot blasting', and the rest of the hull is cleaned of oil, dirt and other contaminants by pressure washing with water and detergent.

The problem with the second method is that it leaves an uneven surface as the old coating is removed in patches and there is a small step change at the point where the blasting occurs. Within the industry studies have shown significant savings in fuel when the hull is prepared to Sa 2½ as opposed to spot blasting. As the IMO require year on year efficiency savings and cost of fuel then full preparation to Sa 2½ might work out to be cost-effective in some cases.

Coating systems

The careful attention to surface preparation is wasted unless backed up by a high quality coating system that is correctly applied. The priming coat is very important as it must adhere to the metal surface. It must be capable of withstanding the wear and tear of everyday working and be easy to apply with airless spray equipment.

The substrate is usually an epoxy resin and the pigment is usually grey. The primer must be compatible with the hard wearing, watertight topcoat. The coatings must be applied on clean, dry surfaces with some of the most modern 'self-polishing' coatings having the ability to make their top surface smother by evening out the imperfections in the hull's surface.

Cathodic protection

If three dissimilar metals are immersed in an electrolyte, the metal lowest on the electrochemical scale becomes the anode, the remaining two being cathodes. Thus if copper and iron are immersed in sea water, they may be protected by a block of zinc which is then known as a *sacrificial anode*, since it is allowed to corrode in preference to the copper and iron. Thus zinc or magnesium anodes may be used to protect the propeller and stern frame assembly of a ship, and will, at the same time, reduce corrosion of the hull due to differences in the steel.

Water ballast tanks may also be protected by sacrificial anodes. It is first essential to remove any rust or scale from the surface and to form a film on the plates which prevents any further corrosion. Both of these functions are performed by booster anodes which have large surface area compared with their volume (e.g. flat discs). These anodes allow swift movement of material to the cathode, thus forming the film. Unfortunately this film is easily removed in service and therefore main anodes are fitted, having large volume compared with surface area (e.g. hemispherical), which are designed to last about 3 years. Protection is only afforded to the whole tank if the electrolyte is in contact with the whole tank. Thus it is necessary when carrying water ballast to press the tank up.

Electrolytic action may occur when two dissimilar metals are in contact above the waterline. Great care must be taken, for instance, when joining an aluminium alloy deckhouse to a steel deck. The traditional method, such as with some of the great ocean liners, was for the steel bar forming the attachment to be galvanised and steel or iron rivets being used through the steel deck, with aluminium rivets to the deckhouse. A coating of barium chromate between the surfaces forms a measure of protection.

The method used on M.S. *Bergensfjord* was most effective, although perhaps costly. Contact between the two materials was prevented by fitting 'neoprene' tape in the joint (Figure 12.7). Galvanised steel bolts were used, but 'neoprene' ferrules were fitted in the bolt holes, opening out to form a washer at the bolt head. The nut was fitted in the inside of the house, and tack welded to the boundary angle to allow the joint to be tightened without removal of the internal lining. The top and bottom of the joints were then filled with a compound known as 'Aranbee' to form a watertight seal.

The modern method is with the use of a welded aluminium/steel structural transition joint. An explosion welding technique is used and as the metals join a natural layer of aluminium oxide forms which in the completed process acts to prevent the galvanic corrosion.

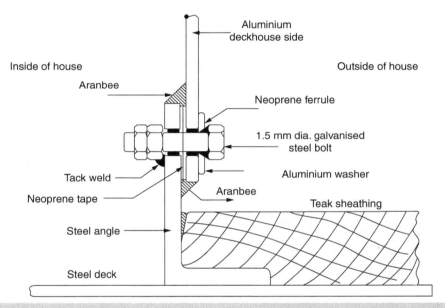

▲ **Figure 12.7** *Connection of aluminium deckhouse to steel deck*

Impressed current system

A more sophisticated method of corrosion control of the outer shell may be achieved by the use of an impressed current. A number of zinc reference anodes are fitted to the hull but insulated from it. It is found that the potential difference between the anode and a fully protected steel hull is about 250 mV. If the measured difference at the electrode exceeds this value, an electric current is passed through a number of long lead–silver alloy anodes attached to, but insulated from, the hull. The protection afforded is more positive than with sacrificial anodes, and it is found that the lead–silver anodes do not erode. A current of 7–350 mA/m^2 is required depending upon the surface protection and the degree to which the protection has broken down.

Design and Maintenance

Corrosion of ships may be considerably reduced if careful attention is paid to the design of the structure. Smooth, clean surfaces are easy to maintain and therefore careful attention should be paid to the welding of ships. If parts of the structure are difficult to inspect, then it is more likely that these parts will not be properly maintained. Efficient drainage of all compartments should be ensured.

Those parts of the structure which are most liable to corrosion should be heavily covered with a suitable coating. Steel plating under wood decks or deck composition is particularly susceptible, for example, as is every tank on the ship. Pools of water lying in plate edges on the deck tend to promote corrosion. If such pools cannot be avoided then the plate edges must be regularly painted. Such difficulties arise with joggled deck plating. An effective inspection system to ensure that all areas are protected and that nothing has been 'missed'.

A warm, damp atmosphere encourages corrosion. Therefore, care must be taken to regularly maintain the structure with respect to deck steam pipes and galley vent trunking.

Reductions in thickness of material of between 5% and 10% may be allowed if the structure is suitably protected against corrosion. If an impressed current system is used for the hull, the maximum interval between dockings/inspection may be increased from 2 to $2\frac{1}{2}$ years.

Maintenance activities

The correct and timely maintenance of the ship's hull and machinery is essential for a number of reasons. The most significant reasons are for the:

- classification society that is ensuring the standard of the vessel
- insurance for the ship
- owner/charter, as the ship needs to be reliable
- regulators, such as the flag state, as the vessel needs to be as environmentally friendly as possible.

The problem is that maintenance is costly and if completed correctly it keeps the vessel operating as it was designed to be. The gives the impression that maintenance is not adding value to the business case for the vessel.

In reality, if the vessel stops or fails to meet its charter commitments, due to the lack of maintenance, then the cost to the business could be considerable. Therefore, the cost of maintenance should be viewed more like an insurance policy and owners/managers cut this cost out at their peril.

Managers of the ship's maintenance can still make a positive contribution to the business of the ship by undertaking the maintenance in the most cost-effective way possible. The most important feature is in the planning of maintenance to take advantage of the natural breaks in the operation of the vessel. For example, undertake maintenance that requires stopping the vessel at the same time as the dry-docking periods, which are enforced downtime and when the vessel is off hire. During this time managers can undertake routine maintenance and survey work as well as the subcontracted work completed by the dry-dock.

Classification societies do realise that both maintenance and the revenue earning opportunity are as equally important and they have developed the Harmonised System of Survey and Certification (HSSC). This is where the time interval between surveys, for different items, can be altered slightly so that a more cost-effective planning process can take place, by combining the work.

Senior technical managers need to work with their commercial colleagues to understand the risks that incorrect or inefficient maintenance will incur.

The costs associated with:

- undertaking the maintenance itself
- spares and the carriage of stock
- standard and specialist tools
- dry-docking
- unplanned downtime due to machinery malfunction
- meeting current and future legislation.

Initially the plan for maintenance will be taken from the manufacturer's recommendations but a more cost-effective method of planning maintenance is condition-based maintenance (CBM). This is where the maintenance is completed when the machinery needs it and not just because the time schedule dictates.

Much more about this subject can be found in Chapter 12 of *Reeds Vol 8*.

Relating to ship's construction, the maintenance to consider is about the hull integrity and its structure. The most reliable method of checking the condition of the hull is to look at it and the most obvious way to accomplish this is to remove the vessel from the water. However, this is expensive, and an in-water survey (IWS) could be used instead of docking the vessel (see pages 182 and 217).

In addition to the condition of the hull, it is vital that the ship's pipework is maintained. The ship's side valves are especially important as it is very difficult to overhaul these with the vessel in the water. Also, students will appreciate that if the ship's side valves are kept is good condition, then the rest of the sea water piping can be repaired at any time, as it can be isolated from the outside sea.

Fouling

The resistance exerted by the water on a ship will be considerably increased if the hull is badly fouled by marine growth. It is found that marine growth will adhere to the ship if the speed is less than about 4 knots. Once attached, however, the growth will continue and will be difficult to remove despite the speed. The type of fouling depends upon the nature of the plant and animal life in the water.

It is essential to reduce fouling, since the increase in resistance in severe cases may be in the order of 30–40%. This is reflected in an increase in fuel consumption to maintain the same speed, or a reduction in speed for the same power.

The main 'anti-fouling' system used to be made up of toxic coatings – usually mercury based. The coating exudes a biocide poison which inhibits the marine growth. Unfortunately, the poison works at all times and poison is being released into the water.

A recent development in the anti-fouling campaign is the introduction of self-polishing copolymers (spc). This is a paint in which the binder and toxins are chemically combined. Water in contact with the hull causes a chemical and physical change on the surface of the coating. When water flows across the surface, the local turbulence removes or *polishes* this top layer. With the introduction of MARPOL V the anti-fouling coatings are no longer allowed to contain toxins and therefore a low adhesion coating is used and any marine growth is washed off when the ship starts to travel through the water.

In addition to the anti-fouling properties, the polishing of the layers produces a very smooth surface and hence considerably reduces the frictional resistance and hence the fuel consumption.

Fouling of the sea inlets may cause problems in engine cooling, while explosions have been caused by such growths in air pipes. When a ship is in graving dock, hull fouling must be removed by scrapers or high pressure water jets. These water jets, with the addition of abrasives such as grit, prove very effective in removing marine growth and may be used while the vessel is afloat.

One method of removing growth is by means of explosive cord. The cord is formed into a diamond-shaped mesh which is hung down from the ship side, attached to a floating line. The cord is energised by means of a controlled electrochemical impulse. The resultant explosion produces pressure waves which pass along the hull, sweeping it clear of marine growth and loose paint. By energising the net in sequential layers, the hull is cleaned quickly but without the excessive energy which would result from a single charge.

Examination in Dry Dock – Class

Dry-docking a ship is a necessary part of ownership. Traditionally this was the only way that the parts of the vessel usually covered by water could be examined thoroughly. Recently however in-water surveys have gained popularity as the equipment becomes more sophisticated and more companies are offering an approved service.

In many companies it is the responsibility of the marine engineers and/or superintendents to make an official inspection of the hull of the ship after entering a graving or floating dry dock, while in other companies it is the responsibility of the deck officers. It is essential on such occasions to make a thorough examination to ensure that all necessary work is carried out bearing in mind that whoever is making the inspection is doing so on behalf of the owner.

The shell plating should be hosed with fresh water and brushed down immediately to remove the salt before the sea water dries. The plating must be carefully checked for distortion, buckling, roughness, corrosion and defective welding. The welded seams and butt joints should also be inspected for cracks.

Until recently inspections of this nature to the sides of the vessel would have to wait until scaffolding or other suitable equipment could be erected so that people could work at a height safely.

The use of cameras mounted on small unmanned flying vehicles or drones is starting to prove extremely cost-effective in this area of work. The initial survey is undertaken quickly and identifies areas that require closer inspection.

For example, the side shell may be slightly damaged due to rubbing against quays, jetties etc. After inspection and repair the plating should be wire brushed or shot blasted and painted. Any sacrificial anodes must be checked and replaced if necessary, taking care not to paint over the surface.

The ship side valves and cocks are examined, glands repacked and greased where necessary or to a planned maintenance programme. All external grids are examined for corrosion and freed from any blockage. If severe wastage has occurred the grid may be built up with welding or replaced. The shell boxes are wire brushed and painted with a suitable coating.

If the double bottom tanks are to be cleaned, the tanks are drained by unscrewing the plugs fitted at the after end of the tank. This allows complete drainage since the ship lies at a slight trim by the stern. It is essential that these plugs should be replaced before undocking and new jointing should always be fitted.

The after end must be examined with particular care. If at any time the ship has grounded, the sternframe may be damaged. It should be carefully inspected for cracks, paying particular attention to the sole piece. In twin screw ships the spectacle frame must be thoroughly examined. The drain plug at the bottom of the rudder is removed to determine whether any water has entered the rudder.

Corrosion on the external surface may be the result of complete wastage of the plate from the inside. The 'wear-down' of the rudder bearing is measured either at the tiller or at the upper gudgeon. Little or no wear-down should be seen if the rudder is supported by a carrier, but if there are measurable differences the bearing surfaces of the carrier should be examined. If no carrier is fitted, appreciable wear-down may necessitate replacing the hard steel bearing pad in the lower gudgeon. The bearing material in the gudgeons must be examined to see that the pintles are not too slack, a clearance of 5 mm being regarded as a maximum. The pintle nuts, together with any form of locking device, must be checked to ensure that they are tight.

Careful examination of the propeller is essential. Pitting may occur near the tips on the driving face and on the whole of the 'low pressure' side due to cavitation. Propeller blades are sometimes damaged by floating debris which is drawn into the propeller stream. Such damage must be made good as it reduces the propeller efficiency, while the performance is improved by polishing the blade surface. If a built propeller is fitted, it is necessary to ensure that the blades are tight and the pitch should be checked at the same time. If appropriate, the stern gland should be carefully repacked and the propeller nut examined for movement. On vessels with a more traditional arrangement, the wear of the tail-shaft should be measured by inserting a wedge between the shaft and the packing. If measurement exceeds about 8 mm the bearing material should be renewed, 10 mm being regarded as an absolute maximum. There should be little or no wear in the more modern 'oil lubricated' stern tube systems. The wear in this type of tail shaft is usually measured by means of a special gauge as the sealing ring does not allow the insertion of a 'wear-down' wedge. The efficiency and safety of the ship depend to a great extent on the care taken in carrying out such an inspection. The classification society is usually present at the dry-docking. Their surveyors are there to help and give advice, checking that all the class rules are followed.

Emergency Repairs to Structure

During a ship's life faults may occur in the structure or in other parts of the ship's fixtures and fittings. Some of these faults are of little importance and are inconvenient rather than dangerous. Other faults, although apparently small, may be the source of significant damage. It is essential that a guided judgement be made of the relative importance of the fault before undertaking repairs.

In Chapter 2 it was explained that the highest longitudinal bending moments usually occur in the section of the ship between about 25% forward and aft of midships. Continuous longitudinal material is provided to maintain the stresses at an acceptable level. If there is a serious reduction in cross-sectional area of this material, then the ship could split in two. Damage to the plating at the fore end or the stern might be of less importance, although flooding could occur. The flooding should be contained by the watertight bulkheads placed at either end of the vessel (Figure 6.1). A crack in a rudder plate is more of an inconvenience, however, due to the added weight of the water, it could place considerable additional stresses on the supporting structures.

If a plate is damaged, several options are available, such as:

1. Replacing the plate.
2. Cutting out and welding any cracks.
3. Filling in any pits with welding.
4. Fitting a welded patch over the fault.
5. Cutting off any loose plating.

The best solution in all cases is to replace the plate or damaged section. However, this may require docking the ship and in an emergency, or on a short-term basis, the other options could be considered, including a temporary 'patch' repair. In the past this 'patch' might have been a cement box but in recent times more up-to-date materials have been used.

If a crack occurs in a plate, then a hole should be drilled at each end of the crack to prevent its propagation. If the plate is in the midship part of the ship, then great care must be taken. Any welding must be carried out by authorised personnel in the presence of a classification society surveyor, to ensure that all the class rules are applied.

The correct weld preparation and welding sequence must be followed. If the plating is of high tensile steel the correct welding rods must be used. Greater damage may be caused by an untrained welder than leaving the crack untreated.

Pitting in a plate may be filled up with weld metal except in the midship region. In this section it is better to clean out the pit, grind the surface smooth and fair with some suitable filling material to prevent an accumulation of water.

A crack in a rudder plate may be patched once the crack has been stopped from progressing by drilling at either end of the crack. Water in a rudder may also increase the load on the rudder carrier and steering gear and, therefore, the water should be drained as soon as possible. The additional load on the steering gear will not affect the ability of the rudder to turn the ship.

Damage to the fore end usually results in distortion of the structure and leakage, especially if pounding has occurred during heavy weather. It may be possible to partially remove the distortion with the aid of hydraulic jacks, in which case the plating may be patched. Otherwise it is probably necessary to fit some sort of containment such as a cement box.

Damage to a bilge keel may prove serious. In this case it is better to cut off any loose material and to taper the material on each side of the damage, taking care to buff off any projecting material or welding with respect to the damage. A replacement bilge keel must be fitted in the presence of a classification society surveyor.

Engine Casing

The main part of the machinery space in a ship lies between the double bottom and the lowest deck. Above this deck is a large vertical trunk known as the engine casing, which extends to the weather deck. In the majority of ships this casing is surrounded by accommodation. An access door is fitted in each side of the casing, leading into the accommodation. In the case of unmanned machinery spaces this door must carry suitable warnings about the entry into such spaces.

At the top of the trunk the funnel and engine room skylight could be fitted. In the older ships these skylights supplied natural light to the engine room; however modern designs cannot contain glass and casings are used as a supplement to the ventilation, with the whole casing then acting as an air trunk.

The volume taken up by the casings is kept as small as possible since, apart from the light and air space, they serve no useful purpose. It is essential, however, that the minimum width and length should be sufficient to allow for removal of the machinery. The whole of the machinery casing is however protected by a class 1 fire bulkhead.

The casings are constructed of relatively thin plating with small vertical angle stiffeners about 750 mm apart. Flat bars may be used or the plating may be swedged. The stiffeners are fitted inside the casing and are therefore continuous (Figure 12.8). Pillars or deep cantilevers are fitted to support the casing sides. Cantilevers are fitted in many ships to dispense with the pillars which interfere with the layout of machinery. The cantilevers are fitted in line with the web frames.

The casing sides with respect to accommodation are insulated to reduce the heat transfer from the engine room. While such transfer would be an advantage in reducing the engine room temperature, the accommodation would be most uncomfortable. A suitable insulant would be glass fibre in slab form since it has high thermal efficiency and is fire resistant. The insulation is fitted inside the casing and is faced with sheet steel cladding.

Two strong beams are fitted at upper deck or bridge deck level to tie the two sides of the casing together. These strong beams are fitted in line with the web frames and are each in the form of two channel bars at adjacent frame spaces, with a shelf plate joining the two channels.

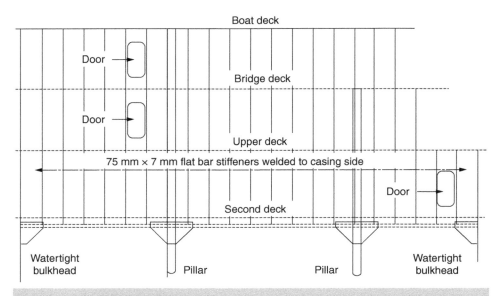

▲ **Figure 12.8** *Elevation of engine casing*

Efficient lifting gear is essential in the engine room to allow the removal of machinery parts for inspection, maintenance and repair. The main equipment is a travelling crane of 5 or 6 tonne lifting capacity on two longitudinal rails which run the full length of the casing. The rails are formed by rolled steel joists, efficiently connected to the ends of the casing or the engine room bulkheads by means of large brackets. Intermediate brackets may be fitted to reduce movement of the rails, as long as they do not obstruct the crane. Figure 12.9 shows a typical arrangement.

The height of the rails depends upon the height and type of the machinery, sufficient clearance being allowed to remove long components such as piston rods and cylinder liners. Large two-stroke crosshead engines will need more space above the engine than will the four-stroke trunk type engines as their components are much shorter. Manufacturers of the larger engines have developed a special 'double jib' crane to reduce the 'empty' space required above the engine for component removal. The width between the rails is arranged to allow the machinery to be removed from the ship. An alternative method used in some ships is to carry two cranes on transverse rails. This reduces the length of the rails but no intermediate support may be fitted.

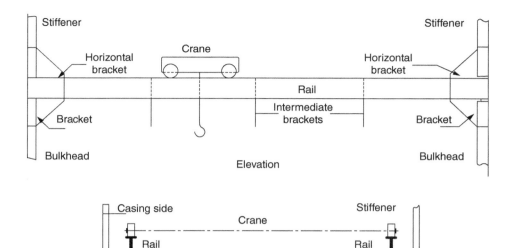

▲ **Figure 12.9** *Central rail support*

Funnel

The size and shape of the funnel depends upon the requirements of the shipowners, builders and designers. At one time tall funnels were fitted to steam ships to obtain the required natural draught and, in passenger ships, to ensure that the smoke and grit were carried clear of the decks. Modern steamships fitted with forced draught boilers did not require such high funnels and the funnel is now a feature of the design of the ship, enhancing the appearance and being a suitable feature for the owners' house markings.

They are sometimes built much larger than necessary, particularly in motor ships where the engine exhaust could be small. They may be circular, elliptical or pear shaped, when seen in a plan view, while there are many varied shapes in side elevation. In many cases the funnel is also designed to house some storage space and/or auxiliary machinery such as ventilating fan units.

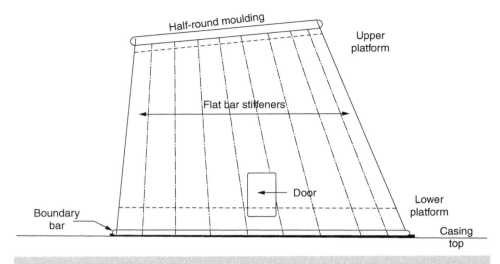

▲ **Figure 12.10** *Funnel construction*

The funnel consists of an outer casing protecting the uptakes. The outer funnel is constructed of steel plate 6–8 mm in thickness. It is stiffened internally by ordinary angles or flat bars fitted vertically, and the exact scantlings depend upon the size and shape of the funnel. The plating is connected to the deck by a boundary angle, while a moulding is fitted round the top to stiffen the free edge.

Steel wire stays might be connected to lugs on the outside of the funnel and to similar lugs on the deck to give extra support especially in bad weather. A rigging screw is fitted to each stay to enable the stays to be tightened. A watertight door is fitted in the funnel, having clips which may be operated from both sides (Figure 12.10).

The uptakes from the boilers, generators and main engine are carried up inside the funnel and are sometime stopped almost level with the top of the funnel (Figure 12.11). A steel platform is fitted at a height of about 1 m inside the funnel. This platform extends right across the funnel, holes being cut in for the uptakes and access. The uptakes are not connected directly to this platform because of possible expansion, but a ring is fitted above and below the plating, with a gap which allows the pipe to slide.

Additional bellows expansion joints are arranged where necessary. At the top a single platform or separate platforms may be fitted to support the uptakes, the latter being connected by means of an angle ring to the platform. In motor ships a silencer must be fitted in the funnel to the main engine exhaust. This unit is supported on a separate

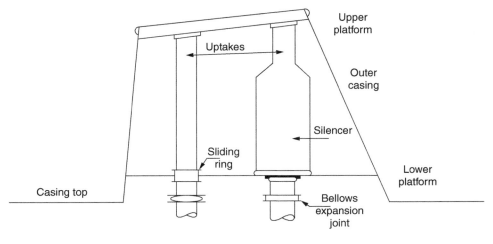

Elevation

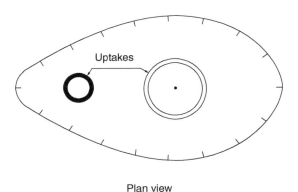

Plan view

▲ **Figure 12.11** *Arrangement of funnel uptakes*

seat. Ladders and gratings are fitted inside the funnel to allow access for inspection and maintenance.

The construction of the exhaust trunking in passenger ships could extend up and point towards the aft end of the vessel. This construction, as mentioned earlier, is to stop any exhaust fumes or debris from being projected across any of the passenger areas. The design of modern ships must incorporate energy efficiency technology. The exhaust gas economiser is a popular choice. This is where boiler feed water is heated by the engine exhaust gas and then fed back into the main boiler where the steam produced is used to provide heating for the ship. Another area of development that affects the design of the funnel is the possible requirement to fit selective catalytic reduction systems to improve the environmental performance of the vessel.

Reducing the environmental footprint of shipping is a significant focus for the industry. Most changes will relate to the propulsion system and the chosen fuel/energy source. More details about these arrangements will be found in the *Reeds Vol 8* and *Reeds Vol 12*.

However, the significant issues relating to the construction of ships will be the design and maintenance of the hull and its coating, as well as the management of ballast water.

SELECTION OF EXAMINATION QUESTIONS – OPERATIONAL/ MANAGEMENT LEVEL

1. Sketch and describe a watertight door. What routine maintenance must be carried out to ensure that the door is always in working order?

2. Draw an outline midship section of a ship and show the position of the following items: (a) sheer strake, (b) garboard strake, (c) stringer, (d) bilge plating, (e) keel plate, (f) floors, (g) frames.

3. With the aid of a sketch showing only the compartments concerned, show the arrangement of windlass and anchor cables. How is the end of the cable secured in the chain locker? What is meant by the terms: (a) hawse pipe, (b) spurling pipe, (c) cable lifter, (d) cable stopper?

4. Sketch and describe the construction of a cruiser stern fitted to a single screw ship and discuss its advantages.

5. Sketch and explain how a large ship is supported in dry dock. Describe the change in the hull stresses that are imposed on a ship resting on the blocks. Detail the precautions that should be taken when refloating the ship in a dry dock.

6. Sketch and describe a transverse section of either an oil tanker or a bulk carrier having two longitudinal bulkheads.

7. Sketch a cargo hatch fitted with steel covers and show how the structural strength of the deck is maintained.

8. Explain what is meant by *longitudinal framing* and *transverse framing*. Which types of ships would have these methods of construction?

9. Explain with the aid of sketches the terms *hogging* and *sagging* with reference to a ship meeting waves having the same length as the vessel. What portions of the structure resist these stresses?

10. Describe with the aid of sketches the terms: (a) camber, (b) sheer, (c) rise of floor, (d) flare. What is the purpose of these?

11. A ship has a small hole below the waterline. What would be the procedure in making a temporary repair around the hole?

12. Show how the hatch and main hold of a refrigerated vessel are insulated. What materials are used? How are the compartments drained?

13 Sketch and describe a stern frame. Show how the frame is attached to the adjoining structure. State the materials used together with their properties.

14. Sketch and describe a deep tank, giving details of the watertight hatch.

15. Sketch and describe a weather deck hatch coaming giving details of the attachment of the half beams. Do the half beams give any strength to the deck?

16. Sketch and describe the different floors used in the construction of a double bottom, indicating where each type is employed. Give details of the attachment of the floors to the adjacent structure.

17. Sketch and describe a transverse watertight bulkhead of a cargo vessel. Show details of the stiffening and the boundary connections.

18. Describe with the aid of a sketch the following types of keel. Show how they are attached to the ship's hull: (a) bilge keel, (b) flat plate keel, (c) duct keel.

19. What are the main functions of: (a) fore peak, (b) after peak, (c) deep tank, (d) double bottom? Give examples of the liquids carried in these tanks.

20. Describe the causes of corrosion in a ship's structure and the methods used to reduce wastage. What parts of the ship are most liable to attack?

21. Sketch and describe a rudder suitable for a ship 120 m long and speed 14 knots. Show how the rudder is supported.

22. Explain the initial examination that would be carried out on the exterior of a ship's hull when in dry dock. State who would complete this examination and the salient points of the inspection necessary to determine the work schedule for the dry dock. Discuss the nature of the defects liable to be found in these areas.

23. Explain with the aid of sketches the arrangement of the following features on a ship: (a) spurling pipe, (b) centre girder, (c) cofferdam, (d) collision bulkhead, (e) podded drive.

24. What precautions must be taken on entering ballast or fuel tanks when empty? Explain why these precautions are necessary.

25. Describe the different methods of ships transporting liquefied petroleum gases (LPG) where the following systems are used: (a) pressurised, (b) fully refrigerated, (c) semi-refrigerated.

26. Sketch and describe the spectacle frame of a twin screw ship. Show how it is attached to the ship.

27. Explain why the plating of the hull and transverse watertight bulkheads are arranged horizontally. Which sections of the ship's structure constitute the strength members, and what design considerations do they receive?

28. Define the following terms: (a) displacement, (b) gross tonnage, (c) net tonnage, (d) deadweight.

29. What are cofferdams and where are they situated? Describe their use in oil tankers.

30. Sketch the panting arrangements at the fore end of a vessel.

31. Explain the following ship conditions and reflect on the type of vessel and/or cargo carried: (a) stiff, (b) tender.

32 Explain why and where transverse bulkheads are fitted in a ship. In which ships are longitudinal bulkheads fitted and what is their purpose?

33. Sketch the construction of a bulbous bow and explain why these are fitted to ships.

34. Sketch and describe an arrangement of funnel uptakes for a motor ship or a steam ship, giving details of the method of support in the outer funnel.

35. Explain the possible reasons for corrosion under each of the following:
    (i)   Connection between aluminium superstructure and steel deck.   (3 marks)
    (ii) In crude oil cargo tanks.   (3 marks)
    (iii) Explain how in each case corrosion can be inhibited.   (4 marks)

36. (i)   Sketch the arrangement of a (bow or stern thruster, showing
          the main power unit, and labelling the principal components.   (4 marks)
    (ii) Explain how this unit operates.   (4 marks)
    (iii) Give reasons for its installation   (2 marks)

37. (i)   Sketch the arrangement of a fin stabiliser unit where the fins retract into a recess in the hull.   (4 marks)

   (ii) Describe how the extension/retraction sequence is carried out.   (4 marks)

   (iii) Define how the action of fin stabilisers effects steering.   (2 marks)

38. (i)   State three different corrosion problems encountered in ship structure.   (3 marks)

   (ii) Define the reasons and effects of each one.   (3 marks)

   (iii) State what precautions are taken to reduce their effects.   (4 marks)

39. Sketch a watertight door giving details of the closing mechanism showing how the watertightness is maintained.   (6 marks)

   Describe the procedure for testing the following for watertightness:

   (i)   A watertight door.   (2 marks)

   (ii) A deep tank bulkhead, and   (2 marks)

   (iii) A hold-bulkhead in a dry cargo ship.   (2 marks)

40. Explain why the 'build up' by welding, patching, cropping or plate replacement is best suited to the following structural defects:

   (i)   Severe pitting at one spot on deck stringer.   (3 marks)

   (ii) External wastage of side plating below scuppers.   (3 marks)

   (iii) Extensive wastage of side plating along waterline.   (4 marks)

SELECTION OF EXAMINATION QUESTIONS – MANAGEMENT LEVEL

1. Sketch and describe the methods used to connect the shell plating to the side frames. How is a deck made watertight where pierced by a side frame?

2. Sketch and describe a hatch fitted to an oil tanker. When is this hatch opened?

3. Discuss the forces acting on a ship when floating and when in dry dock.

4. Explain three possible sources of shipboard vibration. Describe how you would trace the source of the forces causing severe in-service vibrations and the measures that could be taken to reduce the severity of the vibrations.

5. Explain why a tanker is normally assigned a minimum basic freeboard less than that of other types of ship.

6. Sketch and describe a gravity davit. What maintenance is required to ensure its efficient working?

7. Sketch a transverse section through a cargo ship showing the arrangement of the frames and the double bottom.

8. Sketch and describe the arrangements to support stern tubes in a twin screw ship.

9. Explain in detail the forces acting on the fore end of a vessel. Sketch the arrangements made to withstand their effects.

10. Draw a longitudinal section of a dry cargo ship with engines amidships with particular reference to the double bottom, showing the spacing of the floors. What types of ships have no double bottom?

11. Sketch and describe a welded watertight bulkhead indicating plate thickness and stiffener sizes. How is it made watertight? Show how ballast pipes, electric cables and intermediate shafting are taken through the bulkhead.

12. Sketch and describe the arrangement of a rudder stock and gland and the method of suspension of a pintleless rudder. How is the wear-down measured? What prevents the rudder jumping?

13. By what means is the fire risk in passenger accommodation reduced to a minimum? Describe with the aid of diagrammatic sketches the arrangements for fire fighting in the accommodation of a large passenger liner.

14. Sketch the arrangement of a keyless propeller showing how it is fitted to the tailshaft. Discuss the advantages and disadvantages of this design. Explain the method of driving the propeller on to the shaft and how it is locked in position.

15. Explain what is meant by the following terms: (a) exempted space, (b) deductible space (c) net tonnage. Give two examples of (a) and (b).

16. Explain the purposes of a collision bulkhead. Describe with the aid of sketches the construction of a collision bulkhead, paying particular attention to the strength and attachment to the adjacent structure.

17. Describe the effect of and the precautions against the dangers due to water accumulation during fire fighting: (a) while the ship is in dry dock, (b) while the ship is at sea.

18. What precautions are taken before dry-docking a vessel? What precautions are taken before re-flooding the dock? What fire precautions are taken while in dock?

19. Why and where are deep tanks fitted in cargo ships? Describe the arrangements for filling, emptying and drainage.

20. Sketch the following ship stabilisation systems: (a) bilge keel, (b) activated fins, (c) active tanks, (d) passive tank. Explain how each one completes the stabilisation process.

21. Explain the different types of materials that are replacing steel in the building of ships. Describe the vessels concerned and give reasoned explanations why they have replaced mild steel. State the precautions which must be observed when aluminium structures are fastened to steel hulls.

22. Explain with the aid of sketches what is meant by breast hooks and panting beams, giving approximate scantlings. Where are they fitted and what is their purpose?

23. Sketch and describe the freeboard markings on a ship. By what means are they determined? How do the authorities prevent these marks being changed?

24. Sketch and describe the construction of a bulbous bow. Why is such an arrangement fitted?

25. Explain the main causes of corrosion in a ship's internal structure, where this is likely to occur and the measures which can be taken to minimise this action.

26. Draw a cross-section through a modern oil tanker with respect to an oil-tight bulkhead.

    Explain the different methods of generating inert gas for use in a very large crude carrier (VLCC).

27. Sketch the distribution layout of the piping and associated equipment of an inert gas system and explain the safety features incorporated in the system.

28. Sketch and describe the construction of a corrugated bulkhead. What are the advantages and disadvantages of such a bulkhead compared with the normal flat bulkhead?

29. Where do discontinuities occur in the structure of large vessels and how are their effects minimised?

30. Sketch and describe the various types of floors used in a cellular double bottom, and state where they are used.

31. Show how an aluminium superstructure is fastened to a steel deck. Explain why special precautions must be taken and what would happen if no such measures were taken.

32. Sketch and describe briefly: (a) bilge keel, (b) duct keel, (c) chain locker, (d) hawse pipe.

33. Define *hogging* and *sagging*. What members of the vessel are affected by these conditions? State the stresses in these members in each condition.

34. Explain how a bulk carrier is constructed to resist high concentrated loads value.

35. Define the following: gross tonnage, net tonnage, propelling power allowance.

36. Sketch and describe a sternframe. What material is used in its construction and why is this material suitable?

37. Describe with the aid of sketches the arrangements to withstand pounding in a ship.

38. Sketch any two of the following, giving approximate sizes: (a) a peak tank top manhole, (b) a windlass bed, (c) a bilge keel for a large, ocean-going liner, (d) a bollard.

39. Explain the main constructional differences between the following types of vessels: (a) container ship, (b) gas carrier, (c) bulk carrier.

40. Describe the methods adopted in large passenger vessels to prevent the spread of fire. Show how this is accomplished with respect to stairways and lift trunks.

41. Sketch and describe the essential features of ships used for transporting liquefied gas in bulk: (a) free-standing prismatic tanks, (b) membrane tanks, (c) free-standing spherical tanks.

42. A ship suffers stern damage due to collision with a quay. How would the ship be inspected to determine the extent of the damage? If the propeller were damaged state the procedure in fitting the spare propeller.

43. What important factors are involved before new tonnage can be called a *classified* ship?

44. Sketch and describe two types of modern rudders. How are they supported in the ship?

45. Explain the main items to be inspected on the ship's hull during an in-water survey while the vessel is afloat. Explain why a remote controlled unmanned vehicle could be utilised for the examination of the ship's flat bottom.

46. Describe the destructive and non-destructive tests which may be carried out on welding materials or welded joints.

47. Why may tank cleaning be dangerous? State any precautions which should be taken. How does the density of the gas vary throughout the tank during cleaning?

48. A vessel has taken a sudden list in a calm sea. What investigations should be made to ascertain the cause and what steps should be taken to right the vessel?

49. Explain the reasons for each of the following practices:
    (i)   Use of neoprene washers in the connection between aluminium superstructures and ships' main structure.   (4 marks)
    (ii) Attachment of anodic blocks to the underwater surface of a hull.   (3 marks)
    (iii) External shotblasting and priming of hull plating.   (3 marks)

50. Describe the conditions where an 'underwater hull survey' is permitted as an alternative to dry-docking.
    (i)   Explain which parts in particular should be inspected during such a survey.   (5 marks)
    (ii) Describe briefly how such a survey is conducted from a position on board the vessel concerned.   (5 marks)

51. (i)   Sketch the arrangement of a podded drive.   (4 marks)
    (ii) Explain how the unit operates.   (4 marks)
    (iii) Compare the advantages of such units with a more traditional drive and bow/stern thrusters.   (2 marks)

52. Sketch a watertight door showing the frame and closing arrangement and the attachment to the bulkhead. Add in the additional reinforcement carried by the bulkhead to compensate for the aperture.

53. Explain the different contributions made to the reduction of hull resistance by the following:
    (i)   Complete abrasive blasting of hull plating, initially before paint application and during service. (3 marks)
    (ii) Self-polishing underwater copolymer antifouling coatings. (3 marks)
    (iii) Impressed electrical current. (2 marks)
    (iv) Tunnel thrusters. (2 marks)

54. Explain how the 'build up' by welding, patching, cropping or plate replacement is best suited to repairing the following defects:
    (i)   Perforated hollow rudder. (3 marks)
    (ii) Bilge keel partially torn away from hull. (3 marks)
    (iii) Hull pierced, together with heavy indentation of bow below hawse pipe. (4 marks)

55. (i)   Explain the advantages and disadvantages of the different stern arrangements for single and twin screw propulsion. (3 marks)
    (ii) Give reasons for the introduction of the ducted propeller. (4 marks)
    (iii) Explain why the podded drive has an advantage over the more traditional dive systems. (3 marks)

56. (i)   Make a sketch of the essential features of the activating gear fin stabilisers. (5 marks)
    (ii) Explain the operation. (5 marks)

57. Explain why the following conditions can contribute to reduction in ship speed:
    (i)   Damaged propeller blades. (2 marks)
    (ii) Indentation of hull plating. (2 marks)
    (iii) Hole in hollow rudder plating. (2 marks)
    (iv) Ship in ballast. (2 marks)
    (v)   Heavily fouled hull. (2 marks)

INDEX